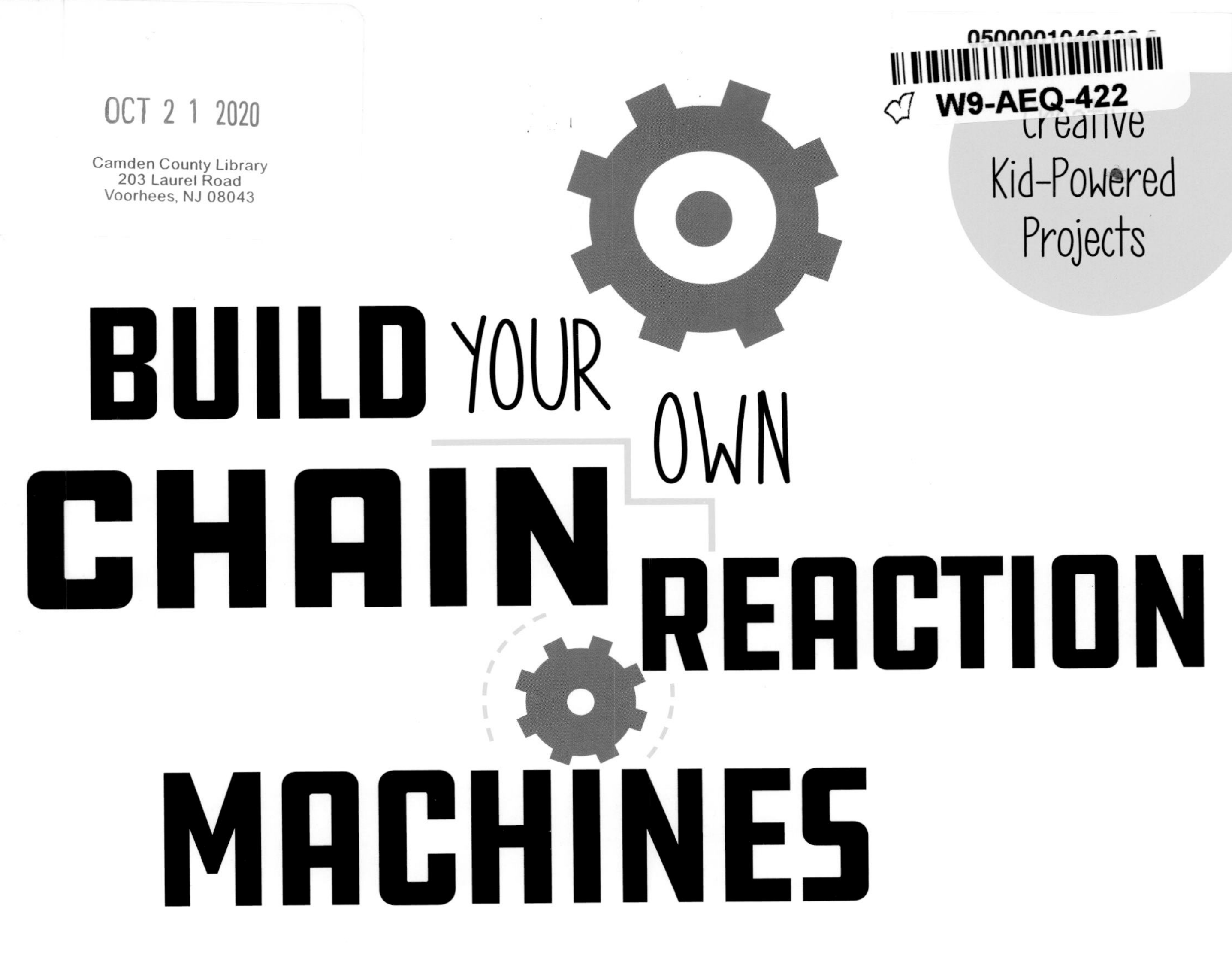


# BUILD YOUR OWN CHAIN REACTION MACHINES

How to Make Crazy Contraptions Using Everyday Stuff

PAUL LONG

Brimming with creative inspiration, how-to projects, and useful information to enrich your everyday life, Quarto Knows is a favorite destination for those pursuing their interests and passions. Visit our site and dig deeper with our books into your area of interest: Quarto Creates, Quarto Cooks, Quarto Homes, Quarto Lives, Quarto Drives, Quarto Explores, Quarto Gifts, or Quarto Kids.



First Published in 2018 by Quarry Books, an imprint of The Quarto Group,
100 Cummings Center, Suite 265-D, Beverly, MA 01915, USA.
T (978) 282-9590 F (978) 283-2742 QuartoKnows.com



Quarry Books titles are also available at discount for retail, wholesale, promotional, and bulk purchase. For details, contact the Special Sales Manager by email at specialsales@quarto.com or by mail at The Quarto Group, Attn: Special Sales Manager, 100 Cummings Center, Suite 265-D, Beverly, MA 01915, USA.

10 9 8 7 6 5 4 3

ISBN: 978-1-63159-526-4

Digital edition published in 2018
eISBN: 978-1-63159-527-1

Library of Congress Cataloging-in-Publication Data

Long, Paul (Paul G.), author.
Build your own chain reaction machines : how to make crazy contraptions using everyday stuff : creative kid-powered projects! / Paul Long.
ISBN 9781631595264 (trade pbk.) | ISBN 9781631595271 (ebook)
1. Machinery--Design and construction--Amateurs' manuals. | 2. Implements, utensils, etc.--Design and construction--Amateurs' manuals.
TJ147 .L66 2018
621.8--dc23
LCCN 2018027954

Design: Laia Albaladejo
Page Layout: Jennifer Giandomenico
Photography: Golbanou Moghaddas
Illustration: Paul Long

Printed in China

For my parents, who taught me a love
of learning and how to take things apart.
And for Golbanou, who follows her passions
and helps me follow mine, no matter what.

# CONTENTS

# INTRODUCTION

I've always been a big fan of books. Fiction, sure, but also DIY and how-to books. As a kid, I got really into origami and crafting, and I always searched the shelves at the library, hoping to find a book that would impart some new and magical knowledge. Nowadays, you can find just about anything on any subject online, especially DIY and how-to videos. And these are great. That's how this book came into being, actually. I created an online course that teaches kids how to invent their own machines using stuff they can find around the house—cardboard, coat hangers, and pencils, held together by hot glue and tape. Things found online are great because they're so easily accessible and the subject matter is broad. But there's something that will always be special about a physical book that you can hold in your hand, whose pages you can turn.

This book is for anyone with curiosity, especially people curious about what makes things tick. It's been said a hundred times, but this is the book I wish I had as a kid. To the best of my knowledge, there's nothing that currently exists that's quite like this book. It's not just another kid's book about making catapults. The chain reaction machines in this book are limited in steps but not in their quality and content. They're laid out step by step, sure, but they're also a bit magical.

Some of these machines take a long time to make. While simple in materials, the mechanisms can be quite complex and the interactions have to work just right in order for everything to come together smoothly. They take patience (something I had little of when I was a kid) and a bit of grit. Often, you might find that your machine won't work on the first, second, or even tenth time. This is normal, even good. It'll just be that much sweeter when it does work. If the words don't make sense, look at the pictures. If you're a visual learner like me, the images will probably be more helpful than the written instructions.

While the materials needed to make the machines in this book are simple, everyday things, don't think you have to use exactly what's listed in the instructions. This is not that type of book. If the instructions say to use a coat hanger and all you have is some wire you found in a drawer, try using that. Everything in this book is a suggestion. If you don't have a pencil long enough to use as a shaft, get creative. The real hope is that this book teaches you the basics, but also inspires. While it allows you to successfully make a machine that can dunk a cookie in milk, I hope it doesn't end there. The point is to take that gained knowledge and apply it to something else, something more creative and more you.

## RUBE GOLDBERG AND CHAIN REACTIONS

Chain reaction machines are also known as Rube Goldberg machines. Rube Goldberg (Reuben Garrett Lucius Goldberg, 1883–1973) was a cartoonist and inventor who drew comics that depicted really roundabout ways of doing ordinary, everyday tasks. At some point, people started making physical versions of these crazy contraptions. There are even Rube Goldberg machine contests. I hope you enter one someday.

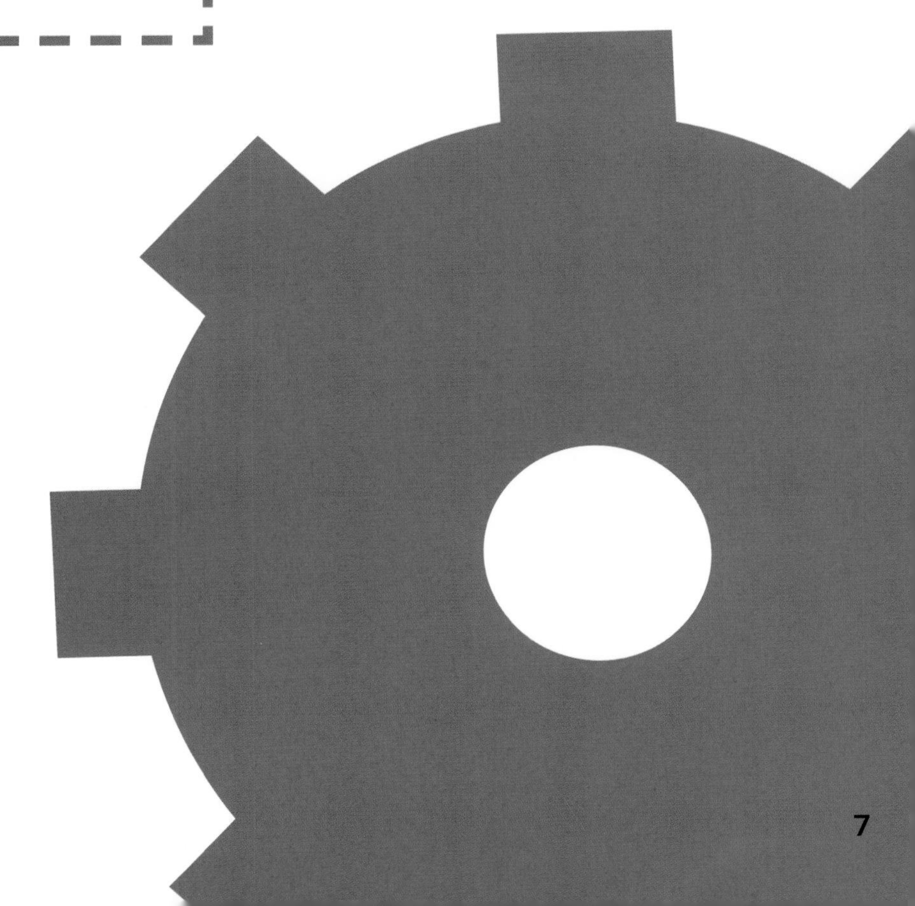

CHAPTER 1

# ESSENTIAL TOOLS, TECHNIQUES AND MECHANISMS

Just as in construction, you need a good foundation before you get started. Here you'll get a glimpse of the best tools to use, along with a few simple but strong tips to give you a head-start on your chain reaction machines.

The Music Maker (see page 102) is one of the nonsensical machines in this book.

# BASIC TOOLS

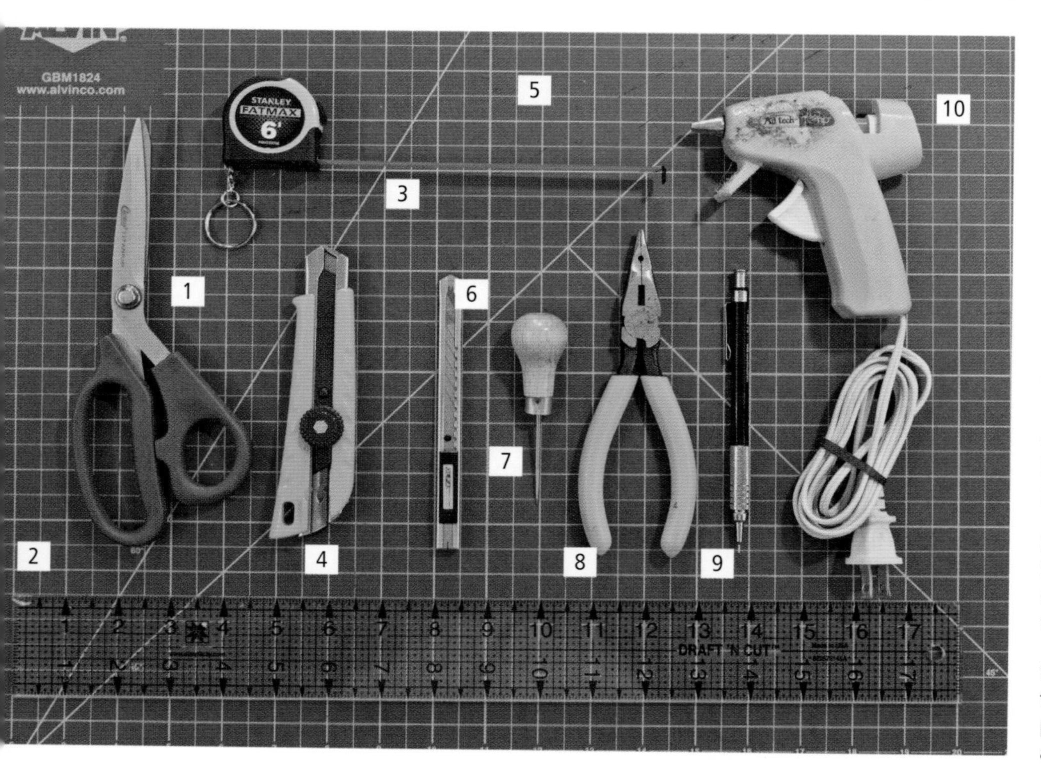


You can make just about anything with cardboard using only a few simple tools. Here's what I use the most:

1. *Scissors:* I normally don't cut cardboard with scissors, but it's great for just about anything else: cereal boxes, cardstock, printer paper, string, fabric, you name it.

2. *Ruler:* The ruler I use is great because it's clear and slightly thick. This makes it perfect to use as a straightedge guide when cutting with the knife. Because the ruler is clear, I can easily cut things to width.

3. *Measuring tape:* For measuring, of course. If you have a ruler, you don't necessarily need this, but it's nice to have.

4. *Utility knife:* I use an Olfa utility knife. The blades are very sharp, and I like the feel of it. This type of knife is best for making long, straight cuts.

5. *Cutting mat:* Cutting mats are fairly cheap and they save you from ruining your kitchen table. The bigger the better, but a 12 x 18" (30.5 x 45.7 cm) works just fine.

6. *Precision knife:* You can use something like an X-acto, but I prefer the Olfa 9 mm knife with the snap-off art blades. They have the same profile as the X-acto, but you can snap them off when they get dull. This knife is best for detailed work, like cutting out circles and other curved cuts.

7. *Awl:* An awl is basically a sharp nail. It's great for poking holes in cardboard when you want to put a toothpick or coat hanger through a piece. I use a cheap leatherworking awl.

8. *Cutting pliers:* These are great for bending and cutting wire. You can also cut toothpicks and other small pieces of wood like craft sticks.

9. *Pencil:* Obviously, you'll need something to make marks with. I like mechanical pencils because the lead stays small and sharp.

10. *Hot glue gun:* Possibly the one tool I couldn't do without. It's not the prettiest glue there is, but it sets super quick, which means you can move on with your making. I like the small, cheap guns best. The smaller sticks heat up fast, they're easier to control, and they don't get as hot, so if you do happen to burn yourself, it won't be too terrible.

# DIY TOOLS

## POKEY PAD

There will be many times in this book when you'll need to poke a hole in your cardboard pieces. If you set your cardboard on a cutting mat and try to poke a hole, it won't go all the way through. You could hold the piece in your hand to make sure you poke all the way through, but then you risk poking your hand. That's where the pokey pad comes in handy. It lets you poke a hole all the way through your piece while it's sitting on a flat surface.

1. Cut a bunch of ½-inch (1.3 cm)-wide strips out of cardboard, approximately 2 inches (5 cm) long **(fig. a)**. Sixteen pieces should be plenty.
2. Glue together the strips to make your pokey pad **(fig. b)**.
3. Now you can poke a clean hole through your entire piece and not poke your hand **(fig. c)**.

FIG. a

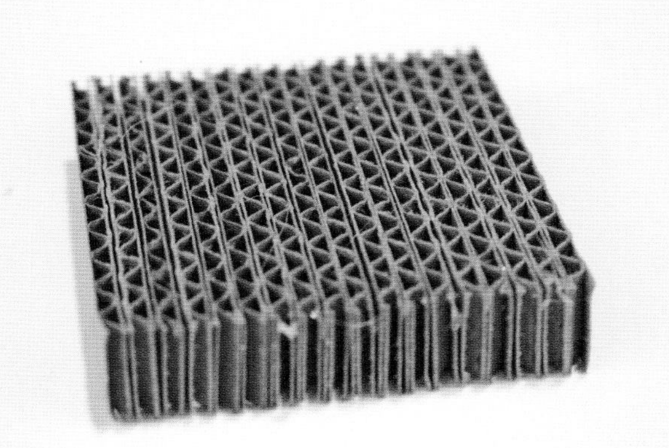

FIG. b

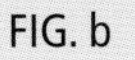

FIG. c

# CENTER FINDER

Normally we'll make our own circles, which means we'll always knows the center. But sometimes, if you need a perfect circle, it's easier to use one that's been made in a factory. Lids and CDs are especially great choices for found circles. The only issue is you don't always know the exact center. This jig, made out of cardboard, will help you find the exact center every time.

1. Cut a piece of cardboard into a perfect square, 4 x 4 inches (10 x 10 cm). The corner of a sheet of paper is perfectly square, so you can use that as a guide. This part is important. If you don't start off with a square, your center finder won't work. Also cut a 1 x 6-inch (2.5 x 15 cm) piece of cardboard, making sure the long edges are straight. Draw guides on your square as shown **(fig. a)**.

2. You want to have an "L" shape as well as a diagonal line from corner to corner. Cut out the "L" shape. Be as careful and as accurate as you can. Keep the small square you just cut to use as a guide **(fig. b)**.

3. Place the square back into the "L" shape so it looks like it was never cut **(fig. c)**. Apply some hot glue to the angled part of the "L" and use the diagonal line as a guide to glue on the long piece of cardboard **(fig. d)**. A craft stick works really well for the long piece.

## BUILD MATERIALS AND TOOLS

- **Two pieces of cardboard:**
  - **1" x 6" (2.5 x 15.2 cm)**
  - **4" x 4" (10.2 x 10.2 cm)**
- **Ruler**
- **Pencil**
- **Utility knife**
- **Hot glue gun**
- **Craft stick (optional)**

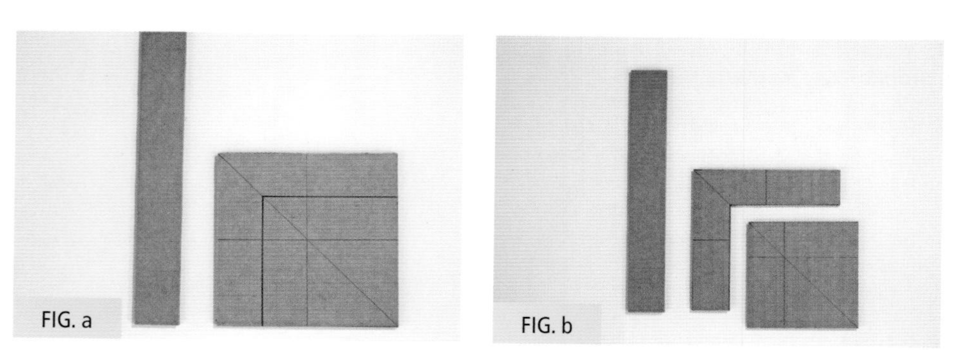

FIG. a

FIG. b

FIG. c

FIG. d

4. You should end up with something that looks like an arrow **(fig. e)**.
5. Place your circle inside the "L" shape, making sure the edges are touching on both sides. Draw a line on the circle using the straight piece as a guide **(fig. f)**.
6. Then rotate your circle approximately 90 degrees and draw another line **(fig. g)**.
7. The point where the two lines intersect is the center of your circle **(fig. h)**. If you want to double-check, you can rotate your circle randomly and draw some more lines. They should all intersect in the same spot. If they don't, then your center finder isn't precise enough.

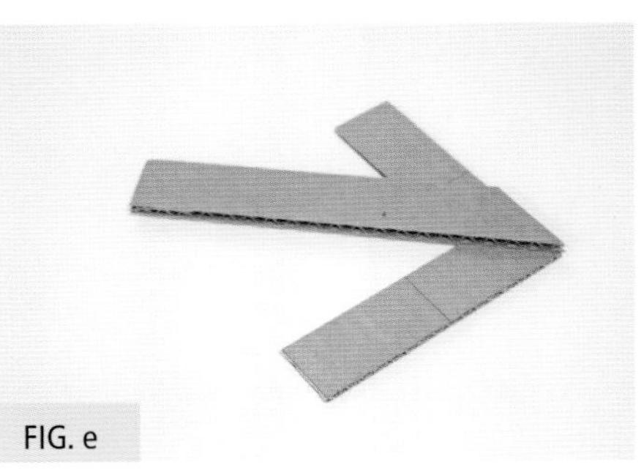

FIG. e

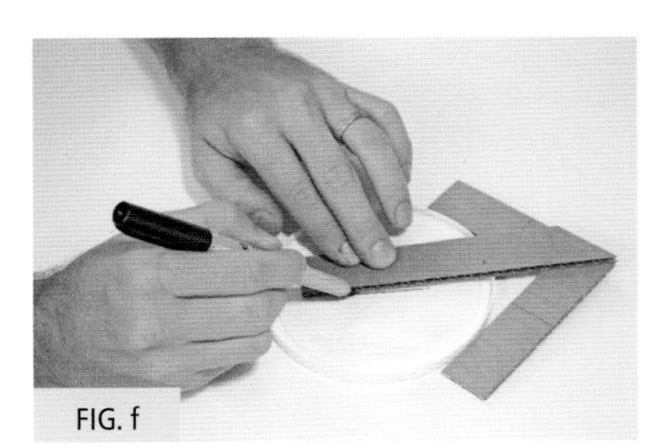

FIG. f

FIG. g

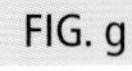

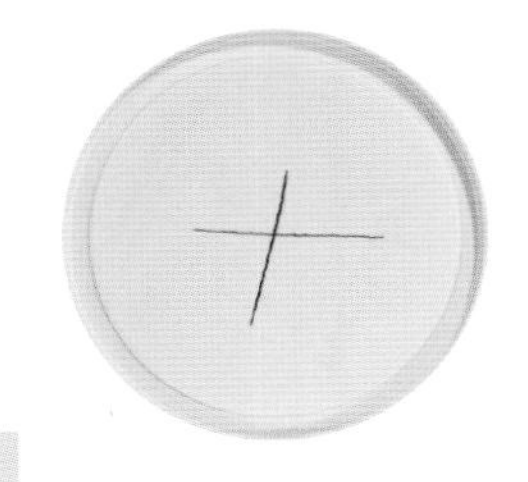

FIG. h

# BASIC TECHNIQUES

## SCORING CARDBOARD

Scoring just means lightly cutting. You're just cutting through the first thin layer of cardboard. You get the best results if you can cut *perpendicular* to the direction of the corrugation. If you score it this way, it makes a nice clean hinge when you fold it on the score line. You can still score it *parallel* to the direction of the corrugation, but sometimes you can get inconsistent results depending on where you scored the corrugated pieces.

## CUTTING COMPLEX SHAPES

It's best to use a cutting mat and straightedge when you're cutting long, straight pieces, but it doesn't work as well when you want to cut out small circles or more curvy shapes. The easiest method I've found is to place your piece on top of a lid or jar, something that has some height with a cutout. Then you can use a small blade to sort of saw up and down, which means you can cut all the way through your piece.

# MAKING THE PERFECT CIRCLE

To make a circle, you first need to know how big you want it. The distance across a circle is called the *diameter*. Half of the diameter is the *radius*. The simplest way to make a circle is to create your own compass. We'll do that by using a strip of thick paper, like a cereal box.

1. Cut the box into a strip that's about ¾ inch (2 cm) wide. If you want a circle that's 2 inches (5 cm) in diameter, make two holes that are 1 inch (2.5 cm) apart **(fig. a)**.
2. Insert a sharp object like an awl or a nail in one hole. This hole is your pivot point and won't move. Place a pencil in the other hole and rotate it around your pivot point. Now you have a 2-inch (5 cm) circle **(fig. b)**.
3. You can make any size circle you want, just by changing the distance from your pivot point to where you place your pencil **(fig. c)**.

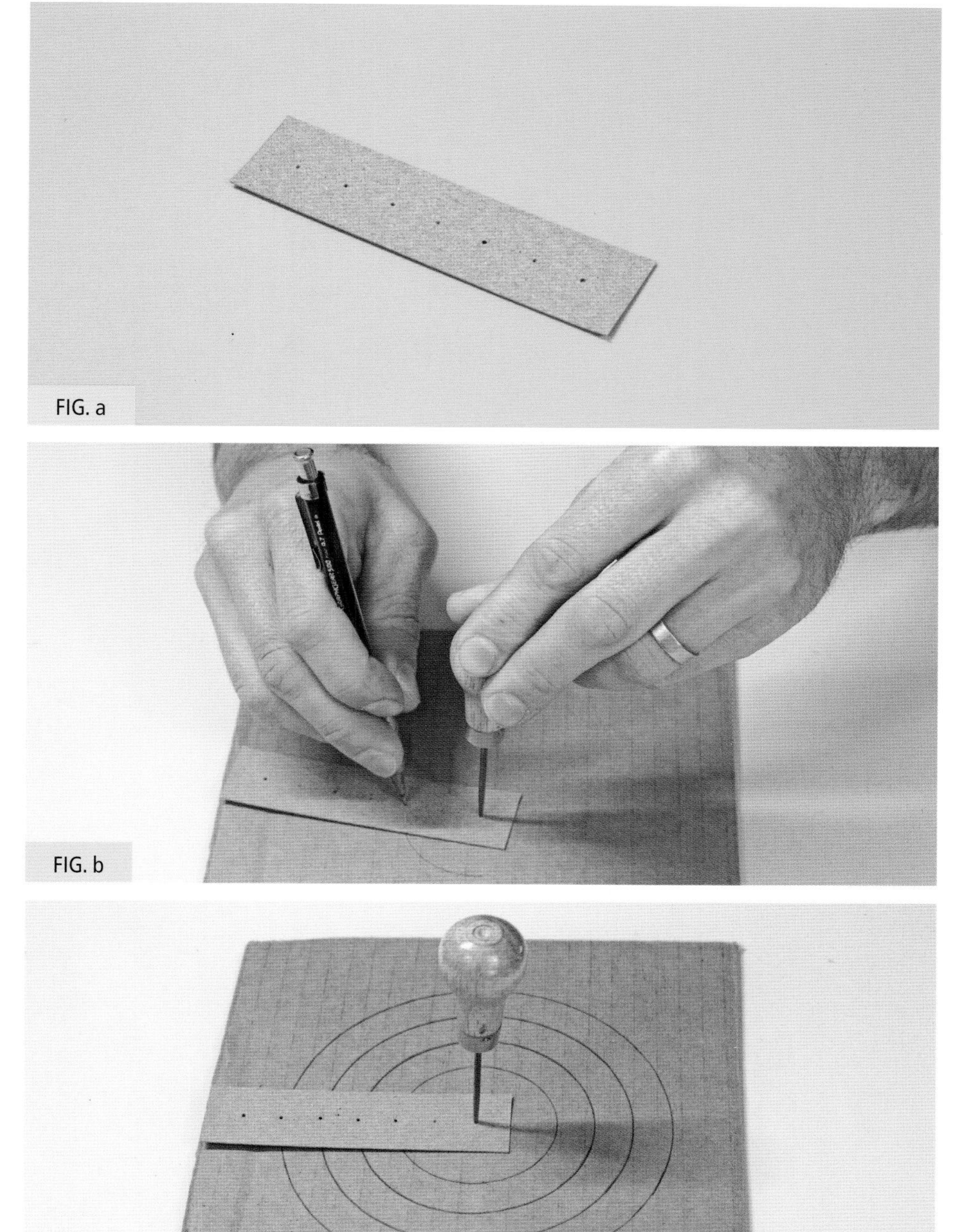

FIG. a

FIG. b

FIG. c

# DIVIDING A CIRCLE

Once you draw your circle, you can use the same paper compass to divide it into pieces (using this method you can divide it into 3 pieces, 6 pieces, 12 pieces, etc.).

1. After you have your circle drawn, poke a hole anywhere on the edge of the circle. This is your first new pivot point **(fig. a)**.
2. Now draw a circle using the same holes (so the circle is the same size). This circle intersects with your original circle. At that intersection point, poke a hole and use that as the pivot point to draw another circle **(fig. b)**.
3. Keep doing this until you have 6 circles. If you draw lines from one hole to another, and go through the center hole, you'll have divided your circle into 6 equal pieces. If you need to divide it into 12 pieces, you can draw a line between the points where the outer circles intersect themselves **(fig. c)**.

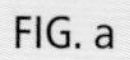

FIG. a

FIG. b

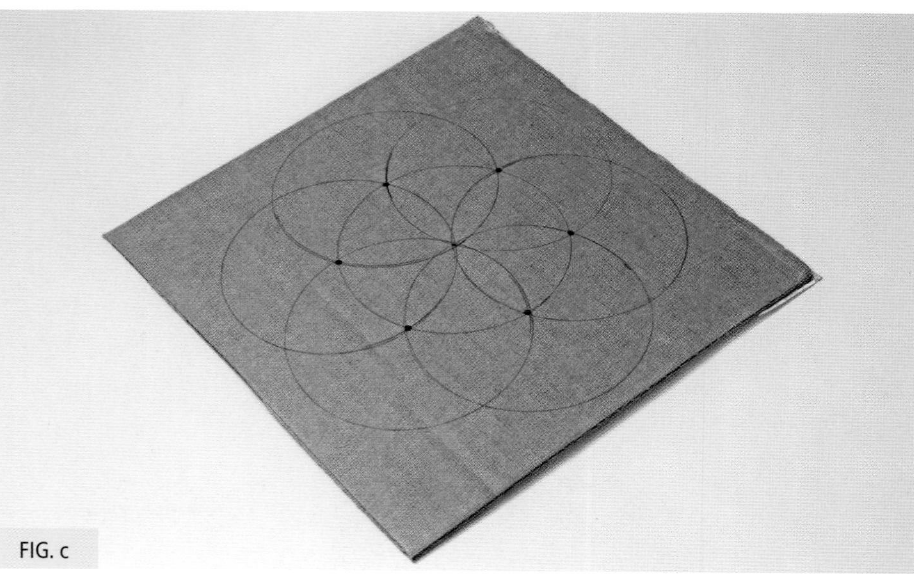

FIG. c

# MAKING A CYLINDER

A cylinder is just two circles wrapped in a thin, flat sheet.

1. I like to use thin cardboard, like a cereal box, for the flat sheet **(fig. a)**.
2. Once you cut your circles, stick a toothpick or skewer through the centers and position them over the thin cardboard. Add a little glue to the thin cardboard and set down your circles. Make sure they're parallel to each other and not crooked. If they're crooked, or your circles are wonky, you won't have a smooth cylinder **(fig. b)**.
3. Let that dry and then run a bead of hot glue about halfway down on both edges and carefully roll the cylinder so the edges of the circles match up with the glue **(fig. c)**.
4. Let that cool and then finish up. If you're making a really big cylinder, you'll have to add glue in smaller sections to make sure it's aligned correctly **(fig. d)**.

FIG. a

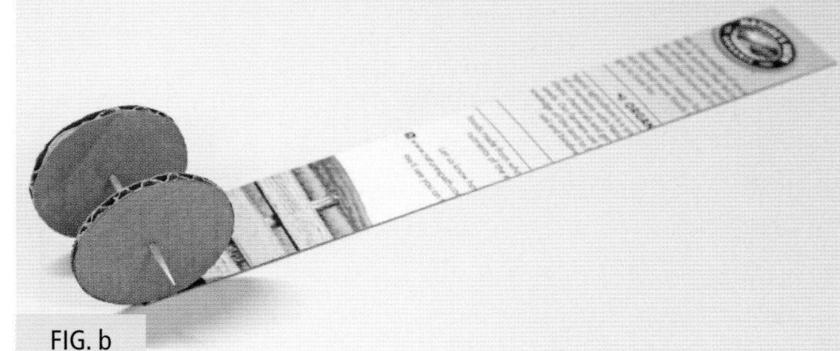

FIG. b

FIG. c

FIG. d

# MAKING A WASHER

A washer can be used as a spacer in between moving parts, or it can seal the end of something that's connected to a shaft. In this case, the washer is keeping the cylinder in place while still allowing it to spin. The simplest washer is a small square of cardboard with a hole in it, but you can make them circular if you want to add a little style. Just make sure the glue is on the shaft (in this case, the toothpick) and the washer, and not on the thing that should be free to spin (that is, the cylinder).

# MAKING A GEAR

A gear is basically a cylinder with teeth all around it.

1. The easiest way to make cardboard gears is to separate the outer layer of cardboard from the corrugated bit **(fig. a)**. Usually one side is easier to remove than the other. Do a test on a scrap piece to see which side comes off easy and clean.
2. Once you have a long section of exposed corrugation, wrap it around your circular pieces to see if the teeth match up **(fig. b)**.
3. In this case, they don't. There's a bit of overlap. This is easy to fix **(fig. c)**.
4. Basically, you just add a thin layer of material to try and make the circle a little bit larger in order to get the teeth spacing correct. You can use the thin piece of cardboard that you just peeled off, or a strip of cereal box might work as well **(fig. d)**.
5. Once you add the paper layer (essentially creating a cylinder), then you can wrap the corrugation around it again to see if the teeth are properly spaced. Once they are, mark where you cut and give it a snip **(fig. e)**.
6. I find it's easier to join the two ends together to make a loop **(fig. f)**, and then slide that onto the cylinder **(fig. g)**.
7. When it's about halfway on, add glue around the inside and push it on the rest of the way **(fig. h)**.

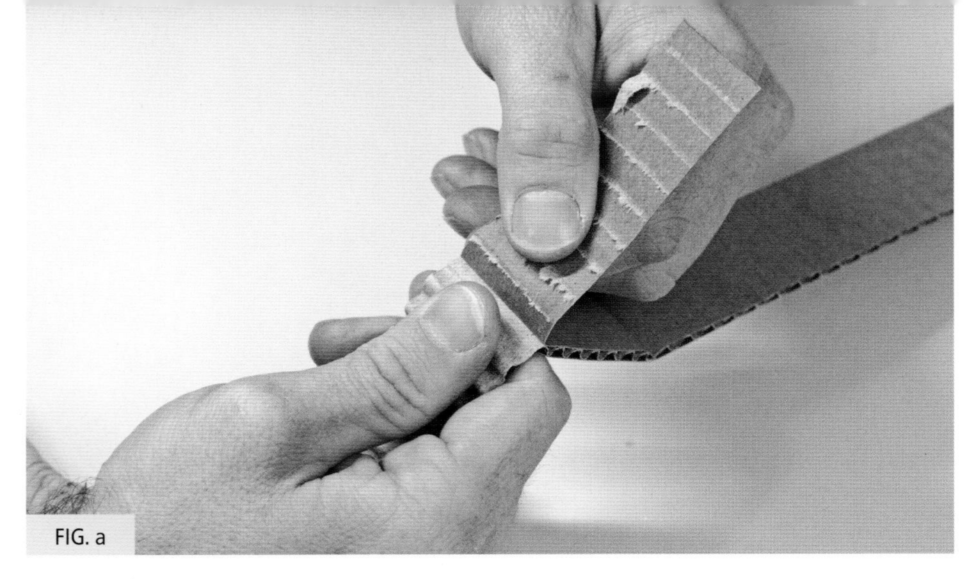
FIG. a

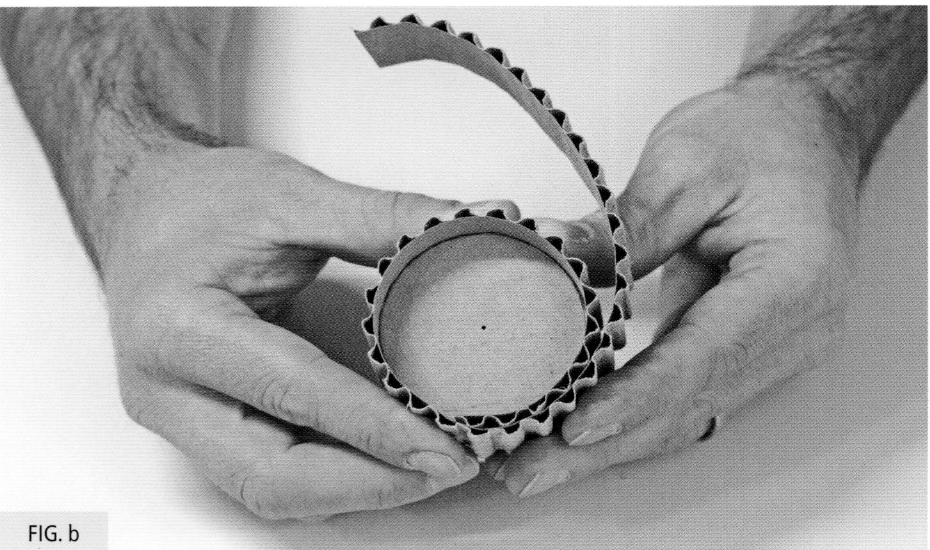
FIG. b

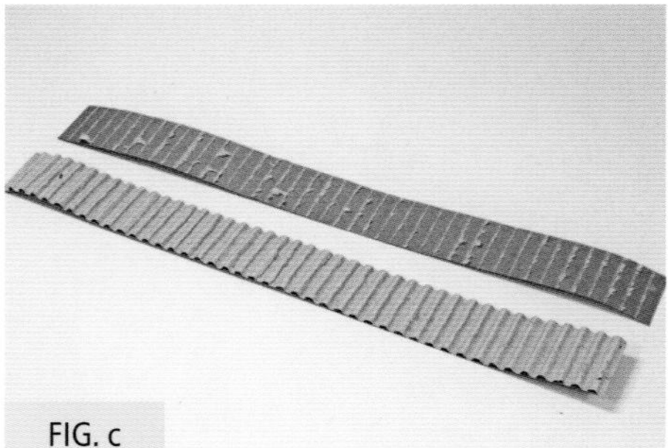
FIG. c

FIG. d

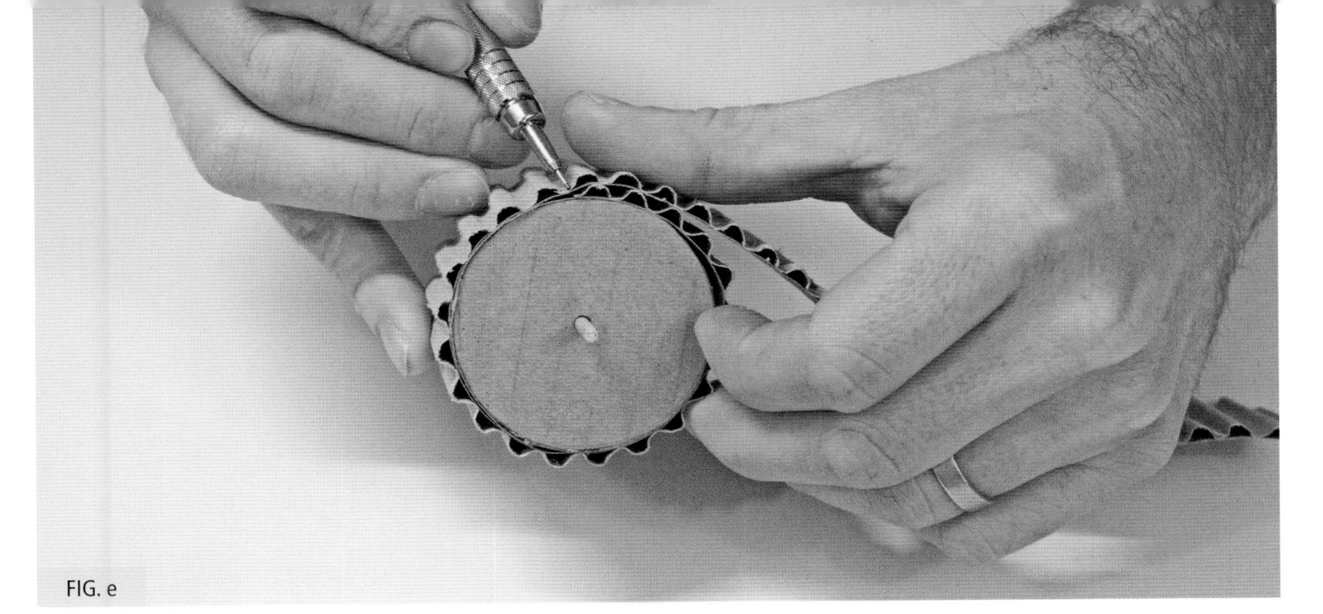
FIG. e

FIG. f

FIG. g

FIG. h

CHAPTER II

# MACHINES FOR YOUR ROOM

These machines are a perfect way to spice up some of the ordinary, everyday things you do in your room. While the outcome is practical, the means for getting there are anything but.

This marble ramp is the first step in activating the Light Switcher (see page 28).

# DOOR KNOCKER

I almost made a knock-knock joke, but I caught myself just in time. This door knocker sounds like a woodpecker with a rubber nose is pecking away, trying to get inside.

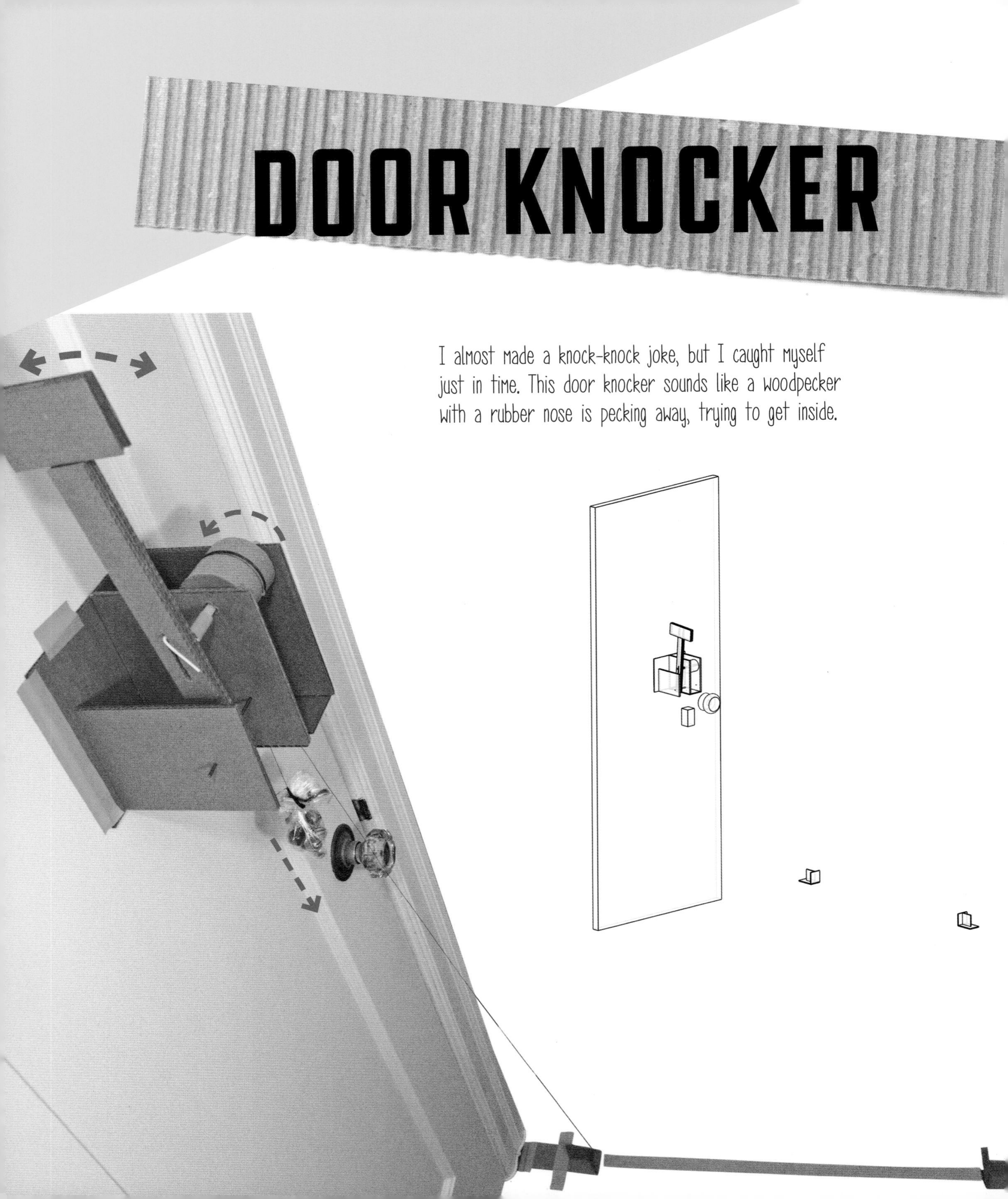

# BUILD MATERIALS & TOOLS

### MAIN STRUCTURE

**CARDBOARD:**

One 6" x 7" (15.2 x 17.8 cm)–**A**
Two 2" x 3" (5 x 7.6 cm)–**B1 & B2**
Two 1" x 2" (2.5 x 5 cm)–**C1 & C2**
Two circles, 2" (5 cm) diameter–**D1 & D2**
Two 1" x 1" (2.5 x 2.5 cm)–**E1 & E2**
Four 5" x 6" (12.5 x 15.2 cm)–**F1, F2, F3 & F4**
Two 5" x 4" (12.7 x 10.2 cm)–**G1 & G2**
One 2½" x 5" (6.4 x 10.2 cm)–**H**
Two 2" x 5" (5 x 10.2 cm)–**I1 & I2**
Four 2" x 2" (5 x 5 cm)–**J1, J2, J3 & J4**
Two 1" x 10" (2.5 x 25.4 cm)–**K1 & K2**
Two 1" x 12" (2.5 x 30.5 cm)–**L1 & L2**
One 2" x 10" (5 x 25.4 cm) thin cardboard–**M**

**OTHER:**

4" (10.2 cm) piece of coat hanger wire with bends at 1½", 2¼", and 3" (3.8, 5.6, and 7.6 cm)–**N**
Toothpick–**O**
4" (10 cm) piece of coat hanger wire–**P**
Small rubber band–**Q**
New pencil or small wooden dowel–**R**
12' (3 m) string–**S**
Bag of marbles (for weight)–**T**

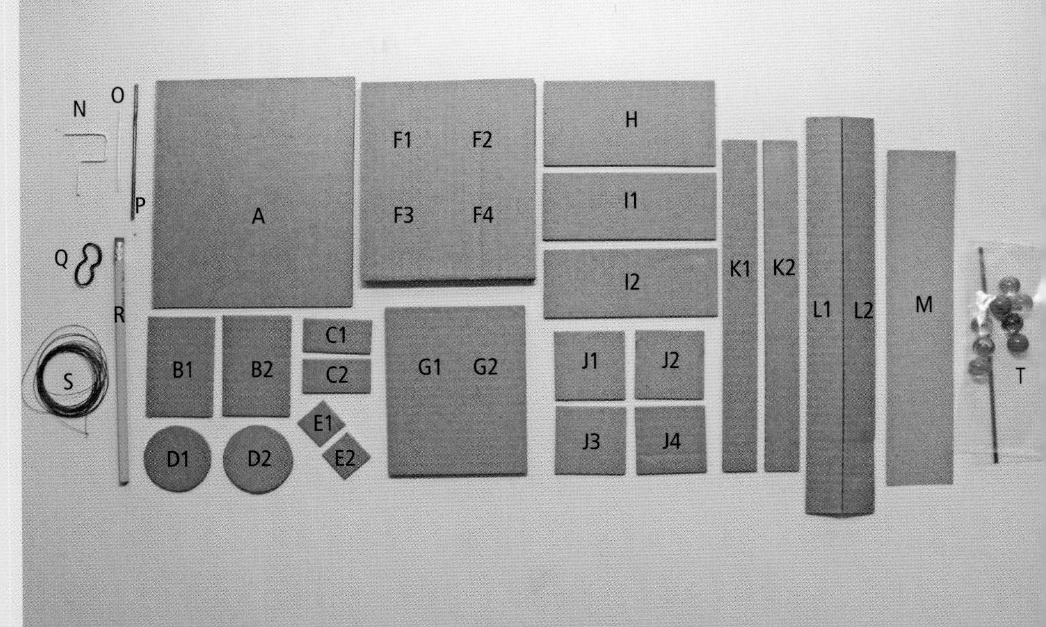


### TOOLS

Glue gun and hot glue
Painter's tape
Sharp pencil
Utility knife
Scissors
Permanent marker
Flat-nose pliers
Awl

# BUILD THE MACHINE

### Make the Hammer Mechanism and Base

1. Pair up F1 and F2 and then F3 and F4. Attach each pair with hot glue so you have two pieces that are doubly thick **(fig. a)**.
2. Glue the doubled-up pieces perpendicularly to A, 2½" (6.5 cm) apart. Leave about ½" (1.3 cm) gap from one edge as shown. Attach H to the edges on one side to form three sides of a box **(fig. b)**.
3. Use M, D1, and D2 to make a cylinder (see page 17) that's 2" in diameter and 2" long (5 x 5 cm). Use a sharpened pencil to enlarge the holes in the center **(fig. c)**.
4. Using a utility knife, cut a fresh pencil or a small wooden dowel to 5" (12.5 cm) long **(fig. d)**.
5. Poke a hole on both of the doubled-up pieces, ½" (1.3 cm) down and 1¼" (3.1 cm) over from the edge. Use a sharpened pencil to enlarge the holes. The trick is to make the holes large enough so the pencil can spin freely. I used the end of my scissors to do the job **(fig. e)**.
6. Insert the cut pencil (or dowel) (R) from the right side, slide it onto the cylinder, and then push it through so about 1½" (3.8 cm) of the pencil sticks out. Take the piece of coat hanger (N) and make sure the wire looks like the one in **fig. f**. Glue the end to the top edge of the pencil. Wrap some tape around the glued bit for extra support.

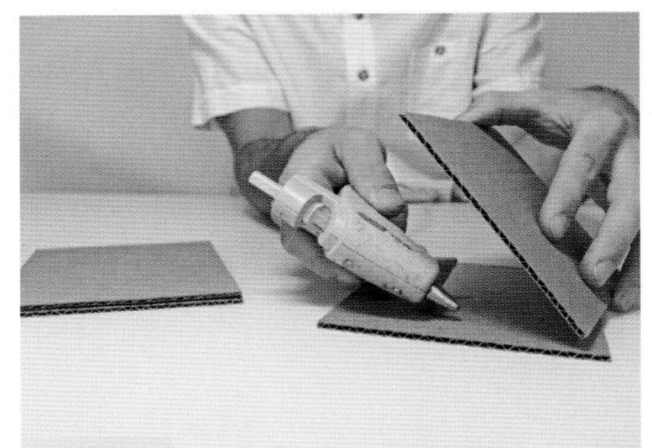

FIG. a

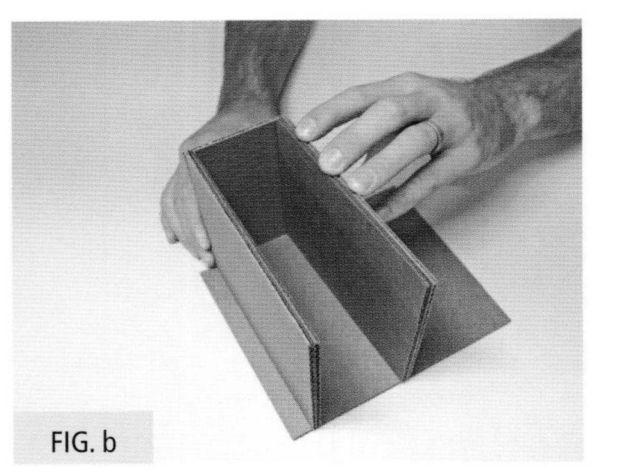

FIG. b

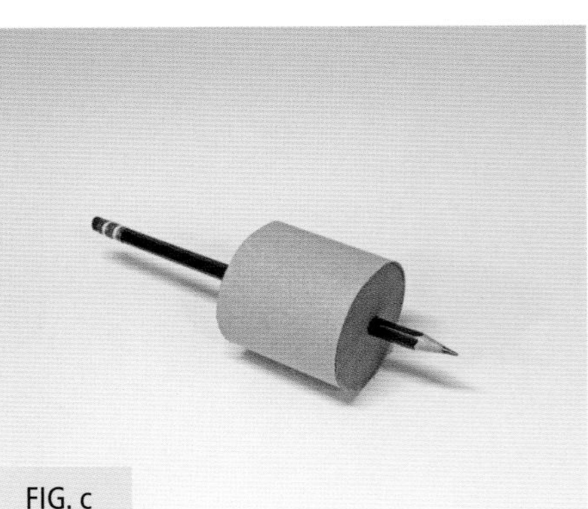

FIG. c

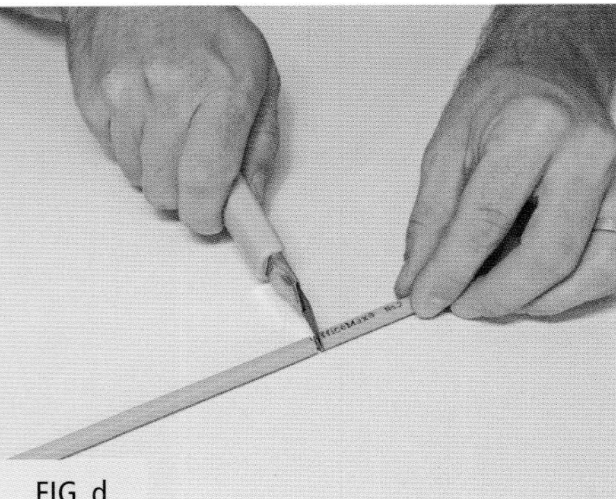

FIG. d

FIG. e

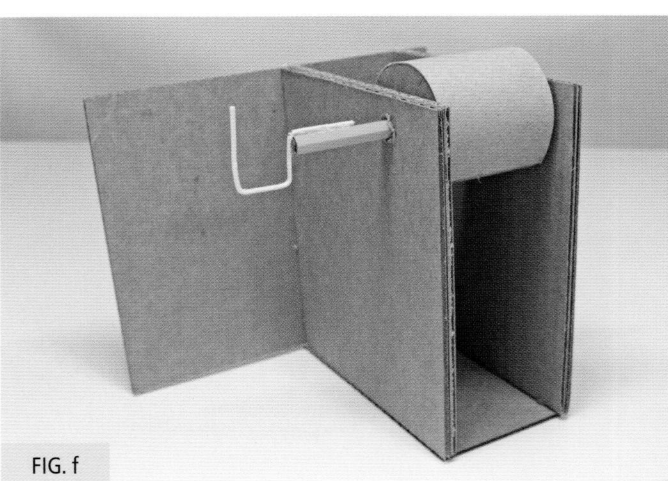

FIG. f

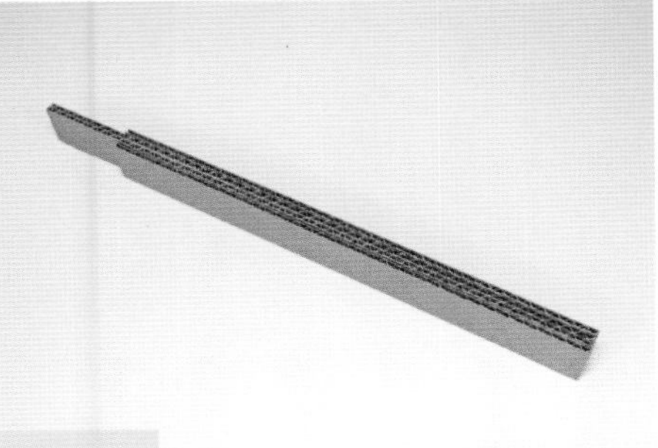
FIG. g

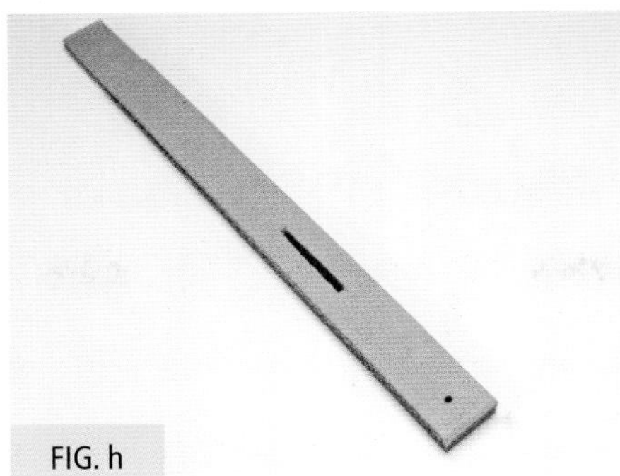
FIG. h

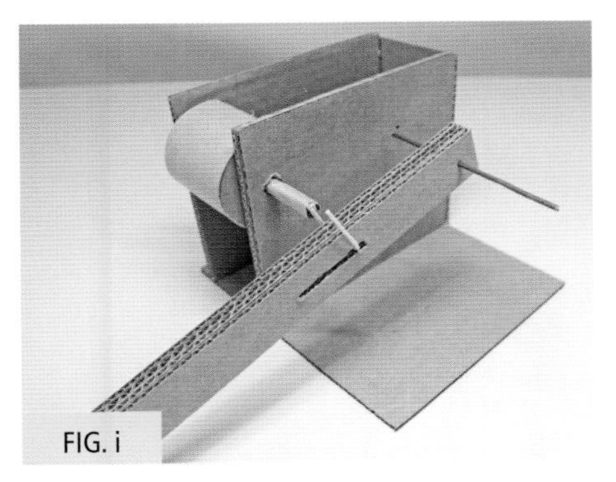
FIG. i

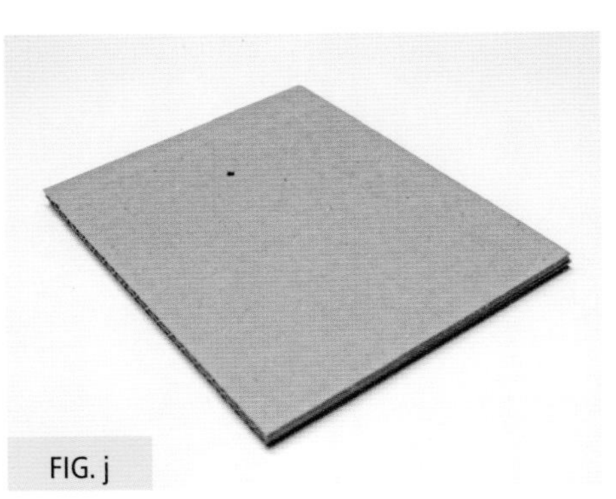
FIG. j

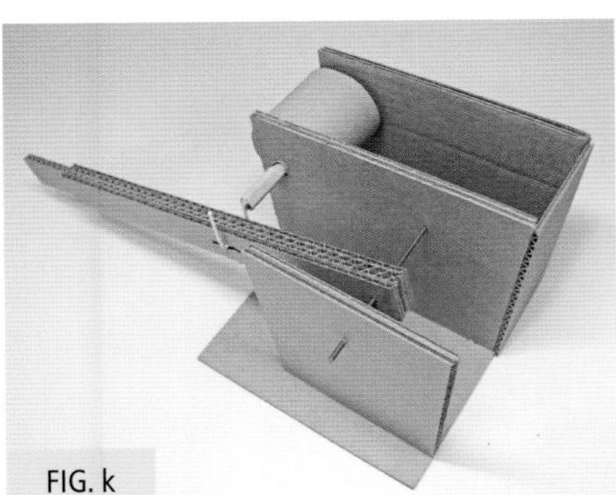
FIG. k

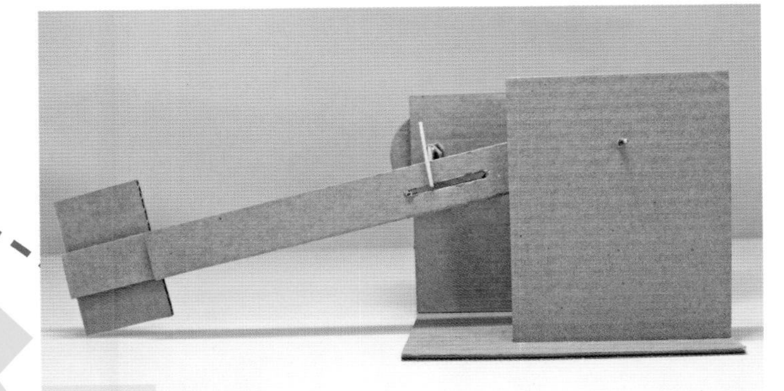
FIG. l

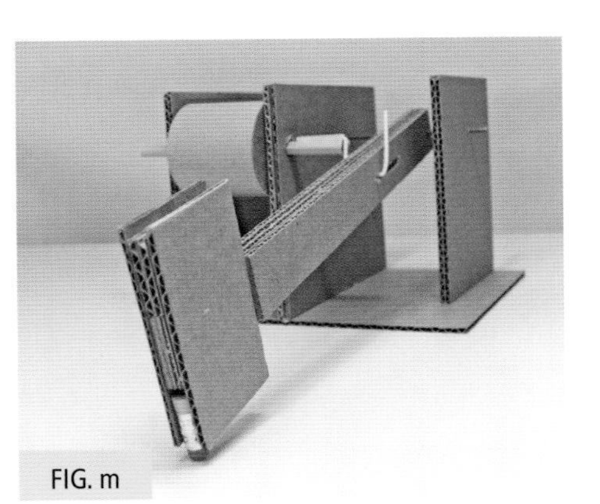
FIG. m

7. Glue L1 and L2 together. Glue K1 and K2 on each side of the doubled-up L1 and L2 so they form a sandwich, with L1/L2 extending out one end **(fig. g)**. This will become the handle of your hammer.

8. With an awl, poke a hole in the thick end of the hammer handle about ½" (1.3 cm) from the bottom. With a utility knife, cut a slit 3" (7.6 cm) from the bottom about 2" (5 cm) long **(fig. h)**. This is where N will go, giving your hammer its back-and-forth action.

9. Slip the hammer handle through the bent coat hanger. Place P through the bottom of your handle where you made the hole, and poke a hole in the box wall 2" (5 cm) from the bottom and 1¼" (3.1 cm) from the edge **(fig. i)**.

10. Glue G1 and G2 together. Poke a hole 1¼" (3.1 cm) from the edge and 2" (5 cm) from the side **(fig. j)**.

11. Slide the other side of P through this hole and glue the cardboard to the large base. Leave a ½" (1.3 cm) gap on the edge **(fig. k)**.

12. Place the bent coat hanger in the "down" position so the hammer handle is as far down as it can go. Glue B1 to the thinner end of the hammer so it's slightly off of the surface, ⅛" (3 mm) or so **(fig. l)**. Repeat on the other side with B2.

13. Sandwich C1 and C2 between the glued B1 and B2 pieces, leaving a gap in the front. Cut off the remainder of the pencil so it just sticks out past the cardboard **(fig. m)**. The eraser will be what knocks on the door.

## Attach the Strings

1. Poke a hole in the cardboard cylinder and glue a long piece of string (48" [122 cm] or so). You can cover the string in tape so it's extra secure **(fig. n)**.

2. Poke a hole in the bottom of the knocker so your string will have a guide **(fig. o)**.

3. Cut a little piece off the toothpick and glue it to the edge of the cylinder like in **fig. p**. Glue E1 and E2 together. Glue this directly below the cylinder, making sure it doesn't interfere with the toothpick. Make sure the corrugation of the cardboard runs parallel with the hammer. We're going to slide a pin through here.

4. Tie another piece of string to the remainder of the toothpick. This piece of string should be super long; its length will depend on how far from your door you want your trip wire. This is going to be your pin, which will keep the cylinder from spinning until it's pulled. Make a cutout for the string **(fig. q)**.

5. Attach J1, J2, J3, and J4 and I1 and I2 as shown in **fig. r.** With the awl, poke a hole in one and add a small loop of wire to the other. These pieces will act as a guide for your trip wire.

6. Rotate the cylinder by hand **(fig. s)**. If the pencil shaft spins inside the cylinder, add a little hot glue by gently pulling I1 and I2 to the side. Make sure you let the glue dry before you let go or it will glue to the other piece as well. We don't want that.

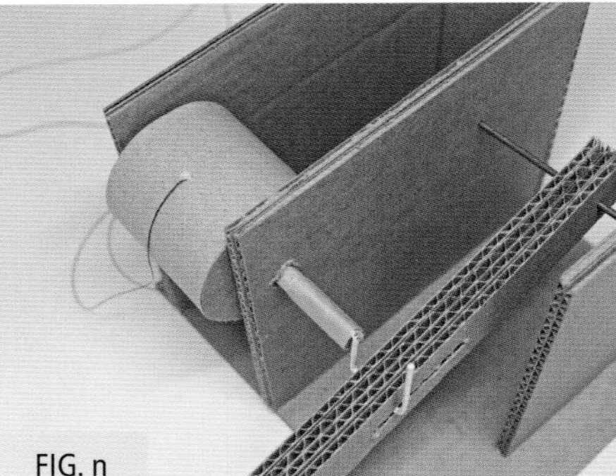

FIG. n

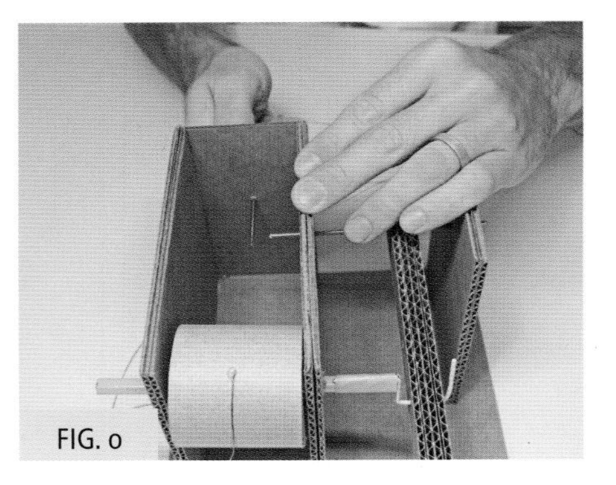

FIG. o

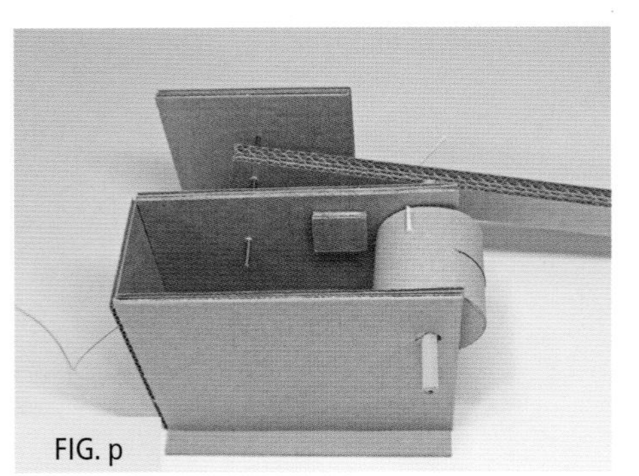

FIG. p

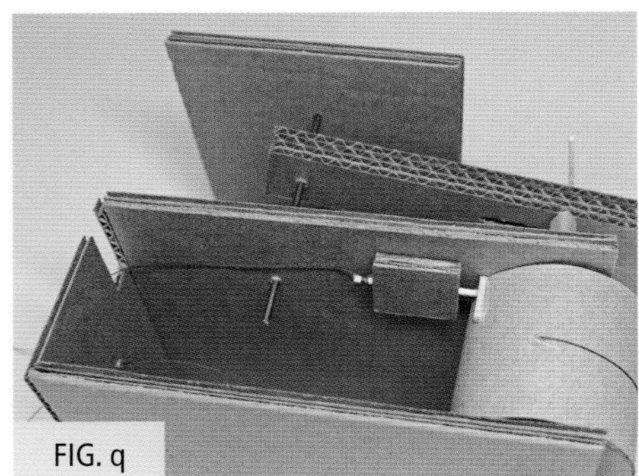

FIG. q

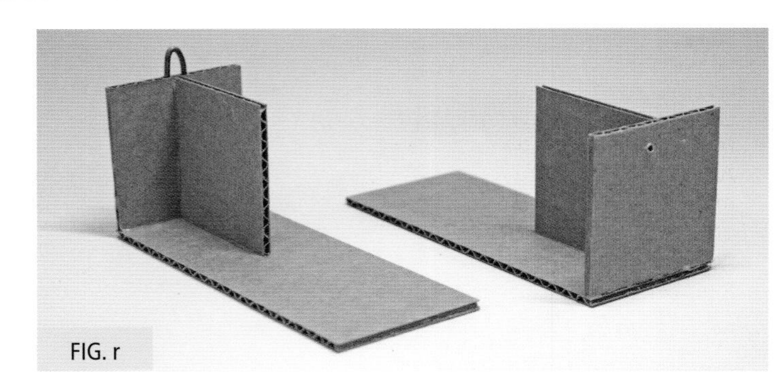

FIG. r

FIG. s

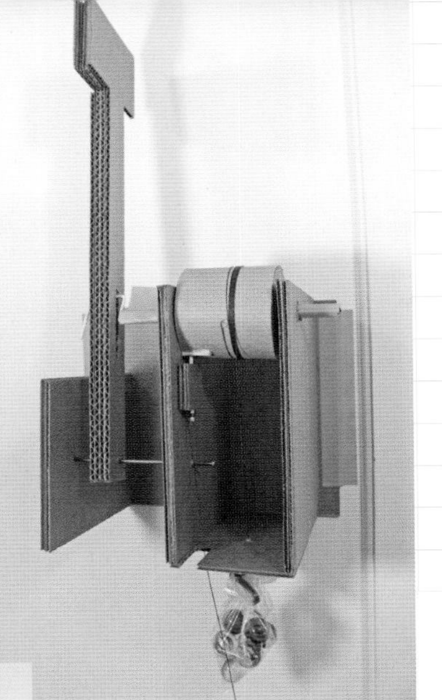
FIG. a

FIG. b

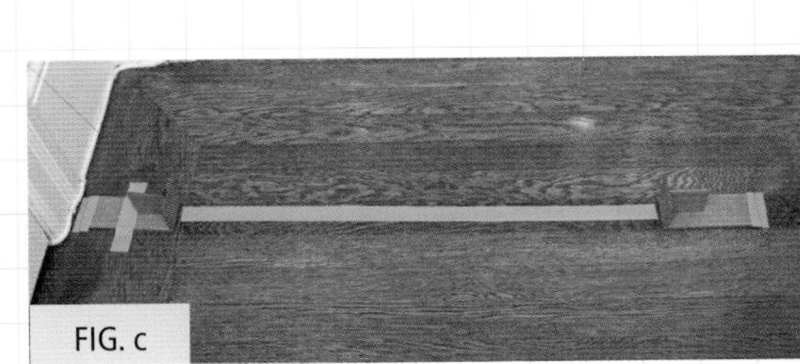
FIG. c

FIG. d

# START THE CHAIN REACTION!

### The Setup

1. Now the fun part! Using painter's tape, tape the door knocker to a door **(fig. a)**. Attach your weight (in my case, a few marbles in a bag) to the string that's attached to the cylinder **(fig. b)**. Then wind up the cylinder. Push up the pin so it rests right behind the toothpick.
2. Tape your trip wire holders to the floor, preferably near your door **(fig. c)**. The holder near the door knocker should have the looped piece. Run your string through the loop and then tie it to the other side. It's hard to see, but the string is running above the long piece of green tape on the floor.
3. Close up of the trip wire string **(fig. d)**.

### Make It Go!

When someone walks by, their feet will go through the trip wire, which will pull out the pin. This allows the weight to fall and the cylinder to spin. The spinning action of the cylinder rotates the bent coat hanger, which makes the hammer go back and forth.

### Troubleshooting

If your hammer gets jammed, it's probably because your pencil eraser is sticking out a bit too far. Trim it back a bit and test it again. If it's still jamming, you might have to trim the head of your hammer. If it's still not working right, the slit you cut in the handle might need to be extended.

### Engineering Know-How

The Door Knocker takes rotational motion and converts it to a back-and-forth motion. This is all because of the bend in the shaft. That bend acts like a cam, which is a sort of protrusion on a rotating shaft. Cams can act as timing mechanisms to push or pull something at a constant pace (assuming the rotation is constant).

# LIGHT SWITCHER

While the Clapper might be a more convenient way to turn off the light, it's not nearly as interesting. You'll save a few pennies on electricity if you employ this machine.

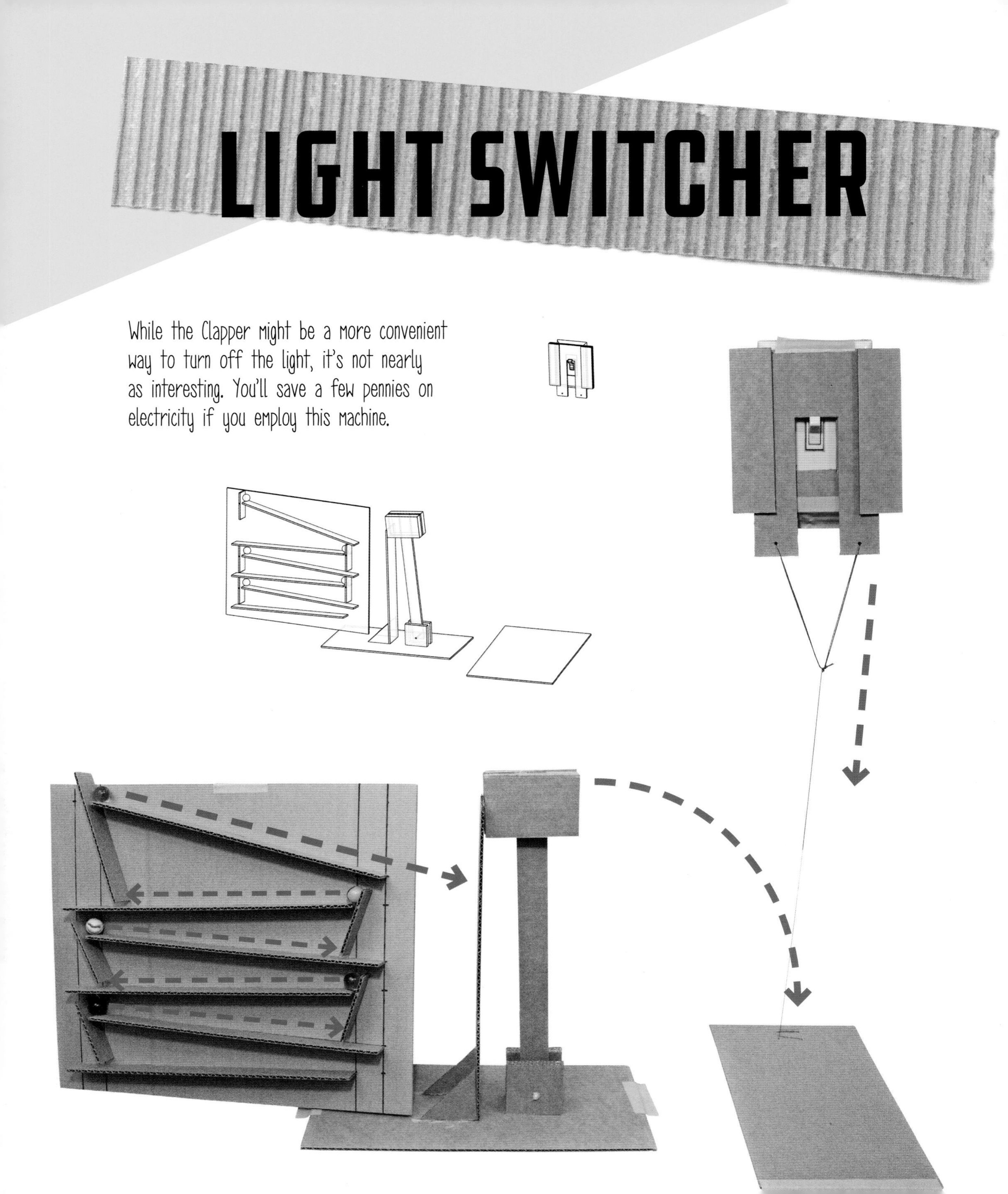

# BUILD MATERIALS & TOOLS

### ZIGZAG RAMP AND SWITCHER MECHANISM

**CARDBOARD:**

One 1" x 5" ( 2.5 x 12.7 cm)–**A**
Four 1" x 2½" (2.5 x 6.4 cm)–**B1, B2, B3 & B4**
One 4" x 4" (10.2 x 10.2 cm)–**C**
Two 1" x 4" (2.5 x 10.2 cm) –**D1 & D2**
Two ½" x 4" (1.3 x 10.2 cm)–**E1 & E2**
Six 1" x 11" (2.5 x 28 cm)–**F1, F2, F3, F4, F5 & F6**
One 12" x 14" (30.5 x 35.6 cm)–**G**
One 6" x 10" (15.2 x 25.4 cm)–**H**
One 3" x 5" (7.6 x 12.7 cm) –**I**

**CARDBOARD PIECE NOT SHOWN:**

One 2" x 12" (5 x 30.5 cm)–**J**

**OTHER:**

6 marbles–**K**
6' (2 m) piece of string–**L**
6 toothpicks–**M**

### HAMMER MECHANISM

**CARDBOARD:**

Two 1" x 12" (2.5 x 30.5 cm)–**N1 & N2**
Seven 2" x 2" (5 x 5 cm)–**O1, O2, O3, O4, O5, O6 & O7**
Ten 2" x 3" (5 x 7.6 cm)–**P1, P2, P3, P4, P5, P6, P7, P8, P9 & P10**
Four 1" x 2" (2.5 x 5 cm)–**Q1, Q2, Q3 & Q4**
Two 1" x 1" (2.5 x 2.5 cm)–**R1 & R2**
One 6" x 12" (15.2 x 30.5 cm)–**S**
One 2" x 11" (5 x 28 cm)–**T**

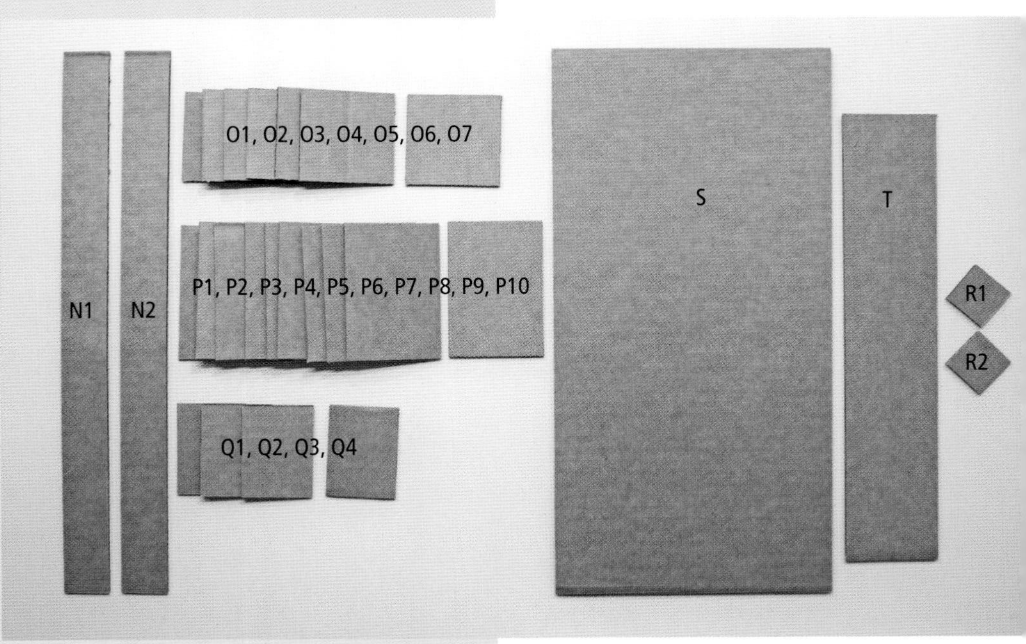


### TOOLS

Marker
Ruler
Glue gun and hot glue
Awl
Scissors or wire cutters
Painter's tape
Utility knife

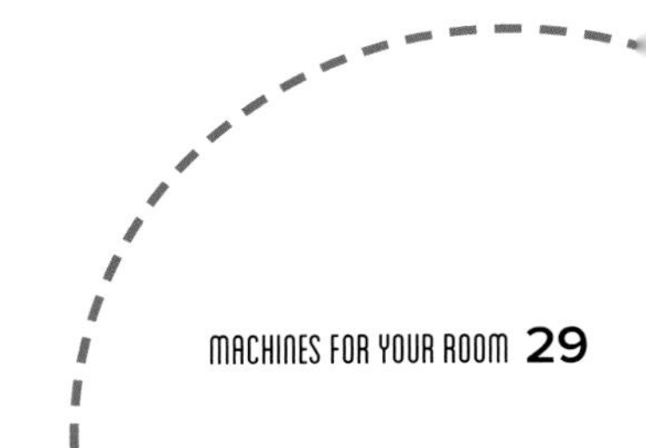

# BUILD THE MACHINE

### Make the Zigzag Ramp

1. On G, draw lines 1" and 2" (2.5 and 5 cm) from the edge, from top to bottom, along the shorter sides. On the left side, make marks at the following intervals: 1", 3", 4", 6", 7", and 11" (2.5, 7.6, 10.2, 15.2, and 17.8 cm). On the right side, make marks at the following intervals: 1½", 2½", 4½", 5½", 7½", and 8½" (3.8, 6.4, 11.4, 14, 19, and 21.6 cm). Use these marks as a guide to draw lines as shown in **fig. a**. These will act as a guide to gluing your marble ramps.
2. Glue F1, F2, F3, F4, F5, and F6 onto the guidelines, perpendicular to the base. With an awl, poke holes next to the higher pieces of the ramp, about ⅛" (3 mm) from the edge, three holes total on each side **(fig. b)**.
3. Insert toothpicks into the poked holes so they're even with the edge of the ramps **(fig. c)**.
4. With scissors or wire cutters, trim the backs of the toothpicks **(fig. d)**, leaving a little sticking out, and add glue to secure them.
5. Insert B1, B2, B3, B4, and A onto the toothpicks, making sure they're able to pivot freely **(fig. e)**.
6. Tape J to the top center back; this will act like a leg of a picture frame so it can stand up by itself **(fig. f)**.

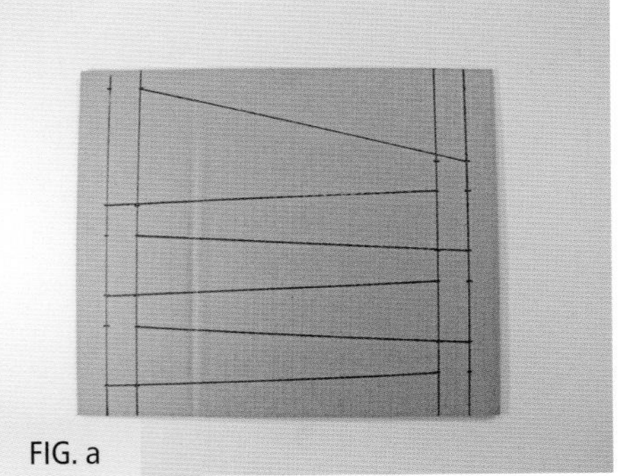

FIG. a

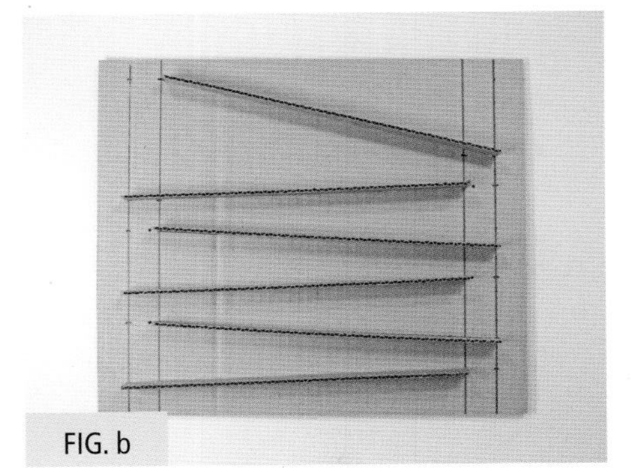

FIG. b

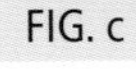

FIG. c

FIG. d

FIG. e

FIG. f

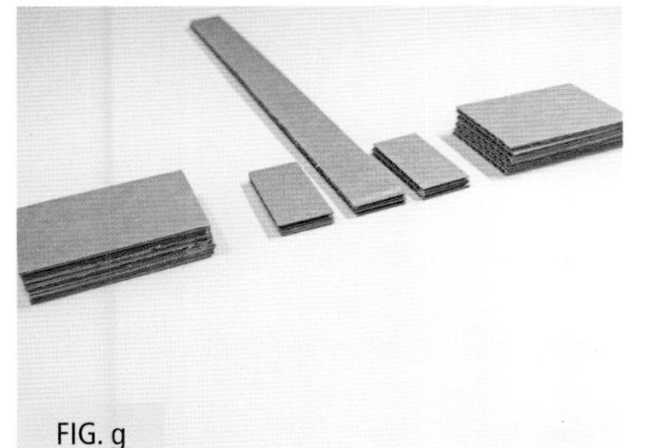

FIG. g

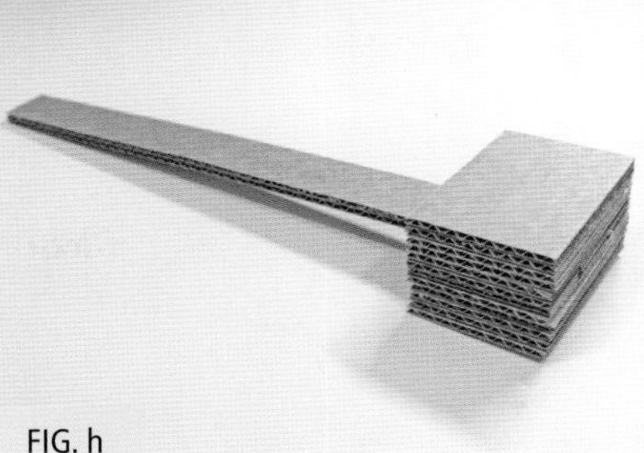

FIG. h

FIG. i

FIG. j

FIG. k

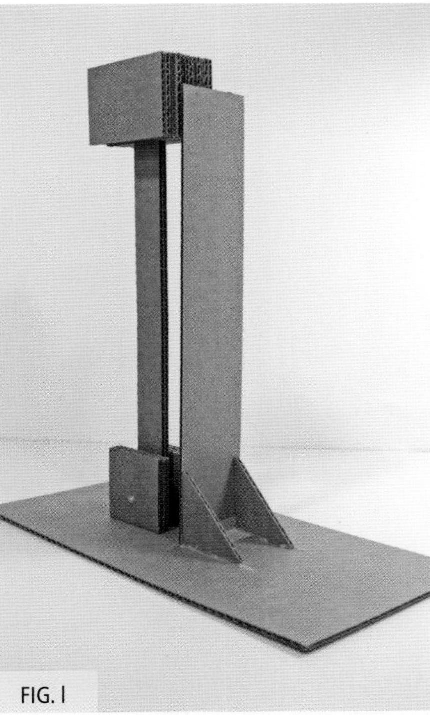

FIG. l

## Make the Hammer Mechanism

1. Next we'll make the hammer, which is what causes the lightswitch to turn off. Glue N1, N2, Q1, Q2, Q3, Q4, P1, P2, P3, P4, P5, P6, P7, P8, P9, and P10 together as shown in **figs. g and h**.

2. Glue R1 and R2 on either side on the bottom of the handle. With an awl, poke a hole through the whole thing, about ½" (1.3 cm) from the bottom **(fig. i)**.

3. Cut O1 in half diagonally. Glue together O2, O3, and O4, and then O5, O6, and O7 so you have two sets of tripled-up pieces **(fig. j)**.

4. Poke holes in the two tripled-up squares, about 1" (2.5 cm) from the bottom. Use the toothpick to create a pivot point for the hammer. Trim the toothpick and add hot glue to the ends **(fig. k)**.

5. Using S as a base, glue the hammer structure about 3" (7.6 cm) from the front. Attach T to S with tape, creating a hinge. It should be about 1" (2.5 cm) directly behind the hammer. Glue the triangles you cut from O1 directly behind this hinged lever **(fig. l)**. Only put glue on the base. Don't glue the side of the triangle to the lever!

6. Once everything is dry, try out the mechanism. One of the marbles will hit the hinged lever right about here (see **fig. m**). The lever will bump the hammer, causing it to fall over. If it doesn't, you can add spacers to the top of the lever to push the hammer a bit farther forward.

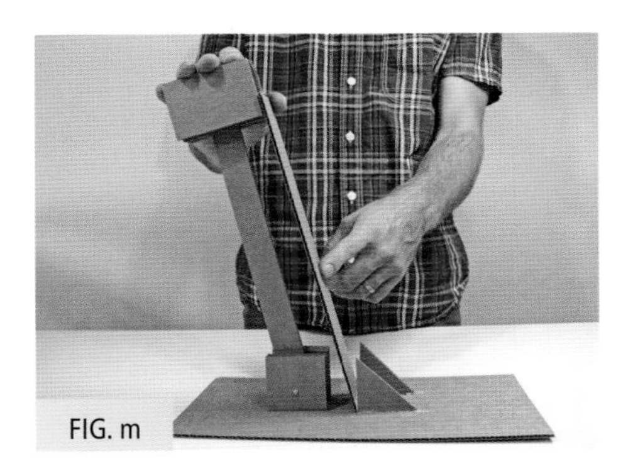

FIG. m

FIG. n

FIG. o

### Build the Light Switcher

1. We need the actual piece that will flip the switch. Glue C, E1 and E2, and D1 and D2 as shown in the **fig. n**. This creates a track for another piece to slide up and down in.

2. Using a utility knife, cut out a 1" x 2" (2.5 x 5 cm) section in the center of the switch piece. Make sure it can slide freely in the switch piece. Cut out a 1" x 3" (2.5 x 7.6 cm) section in I, as shown in **fig. o**, and poke two holes at the bottom. Loop the piece of string through the holes and tie it well.

3. The hammer won't fall on the string directly, so we'll need an intermediate step. A lever like this will do nicely. Poke two holes near the end of H and add slits **(fig. p)**. This will make it easy to adjust our string so we can dial it in just right.

4. Using painter's tape, tape the switch piece piece to your lightswitch **(fig. q)**.

FIG. p

FIG. q

# START THE CHAIN REACTION!

## The Setup

1. Slide the U-shaped piece into the track in the orientation shown in **fig. a**. It will rest on the top of the switch **(fig. b)**. Tie a piece of string to the knotted piece of string. It should be long enough to reach the floor and have some extra length for adjusting.

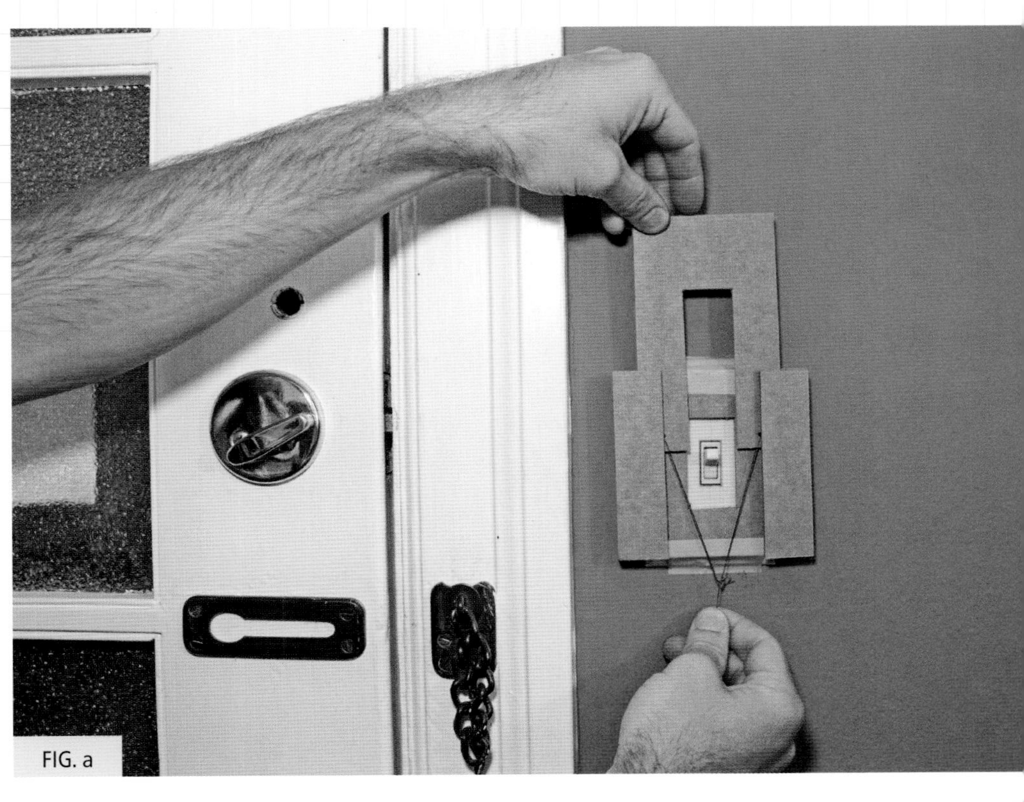

FIG. a

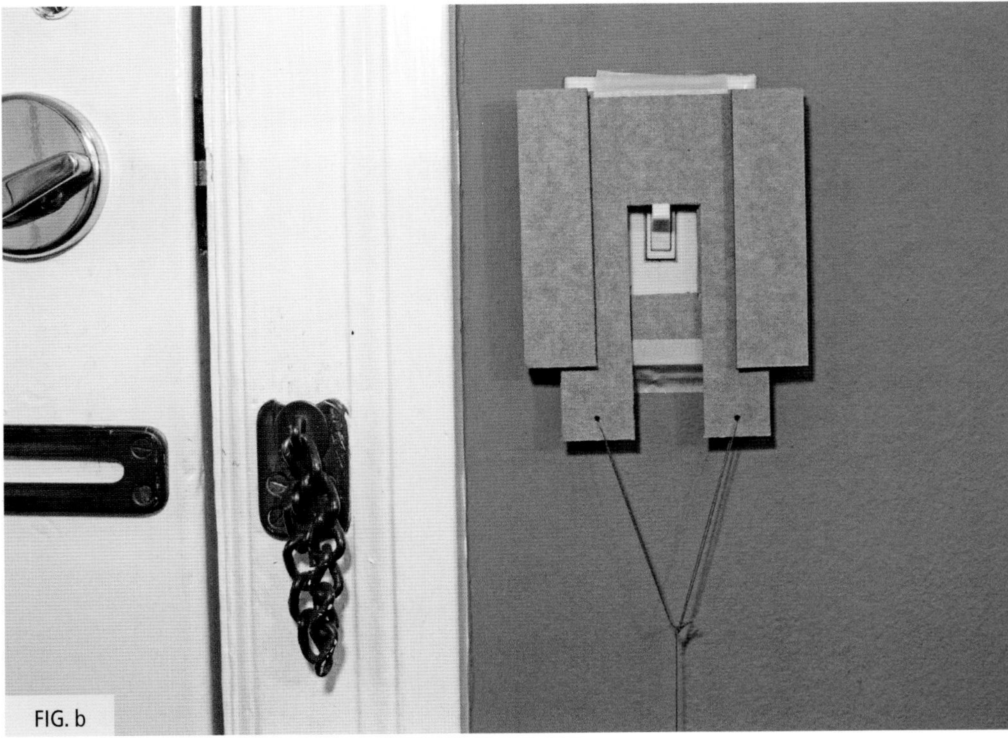

FIG. b

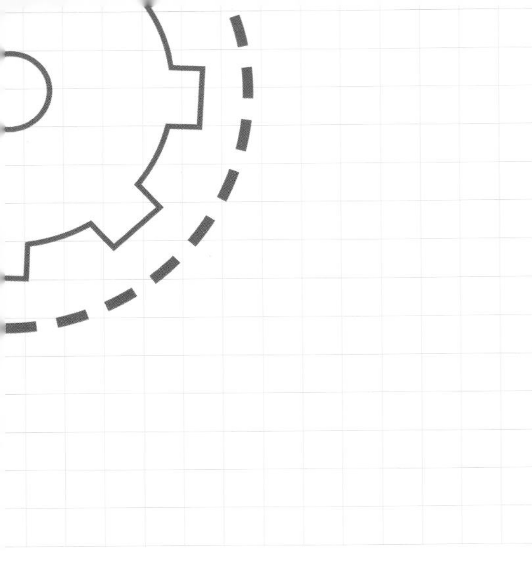

FIG. c

2. Tape the large lever to the floor with the holes and slits facing the wall. Bring your string down and attach it to the lever so it's about 1" (2.5 cm) or so off of the floor **(fig. c)**.

3. Tape the hammer mechanism to the floor, making sure the hammer hits the lever close to the string when it falls over **(fig. d)**.

4. Add the zigzag marble ramp, careful to line it up with the hammer lever. The last marble to roll down should bump this lever, causing the hammer to fall over. Add the other marbles as shown in **fig. e**.

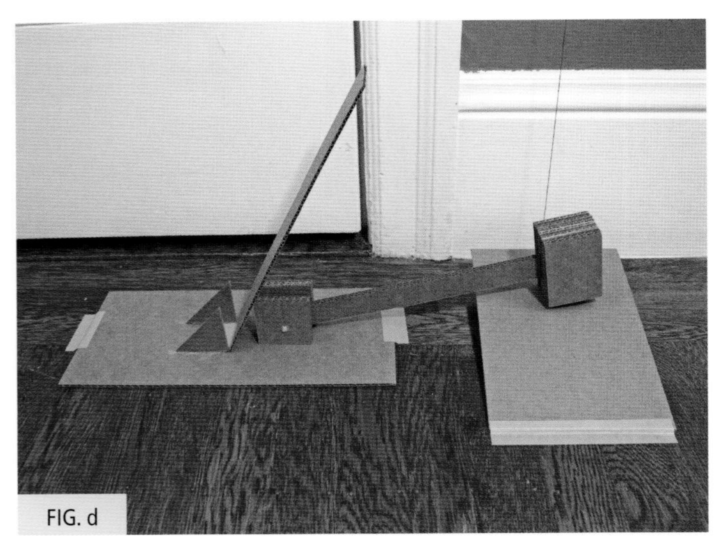

FIG. d

FIG. e

### Make It Go!

To set off the chain reaction you just need to release a marble on the bottom right ramp. It will roll down, bump the small lever, releasing the marble above it. In turn, this marble will roll down its own ramp, bumping the lever on the right. The last marble to be released is at the top of the steep ramp. This allows it to gain more speed and momentum, which will help it push over the hammer, pulling on the string and turning off the switch.

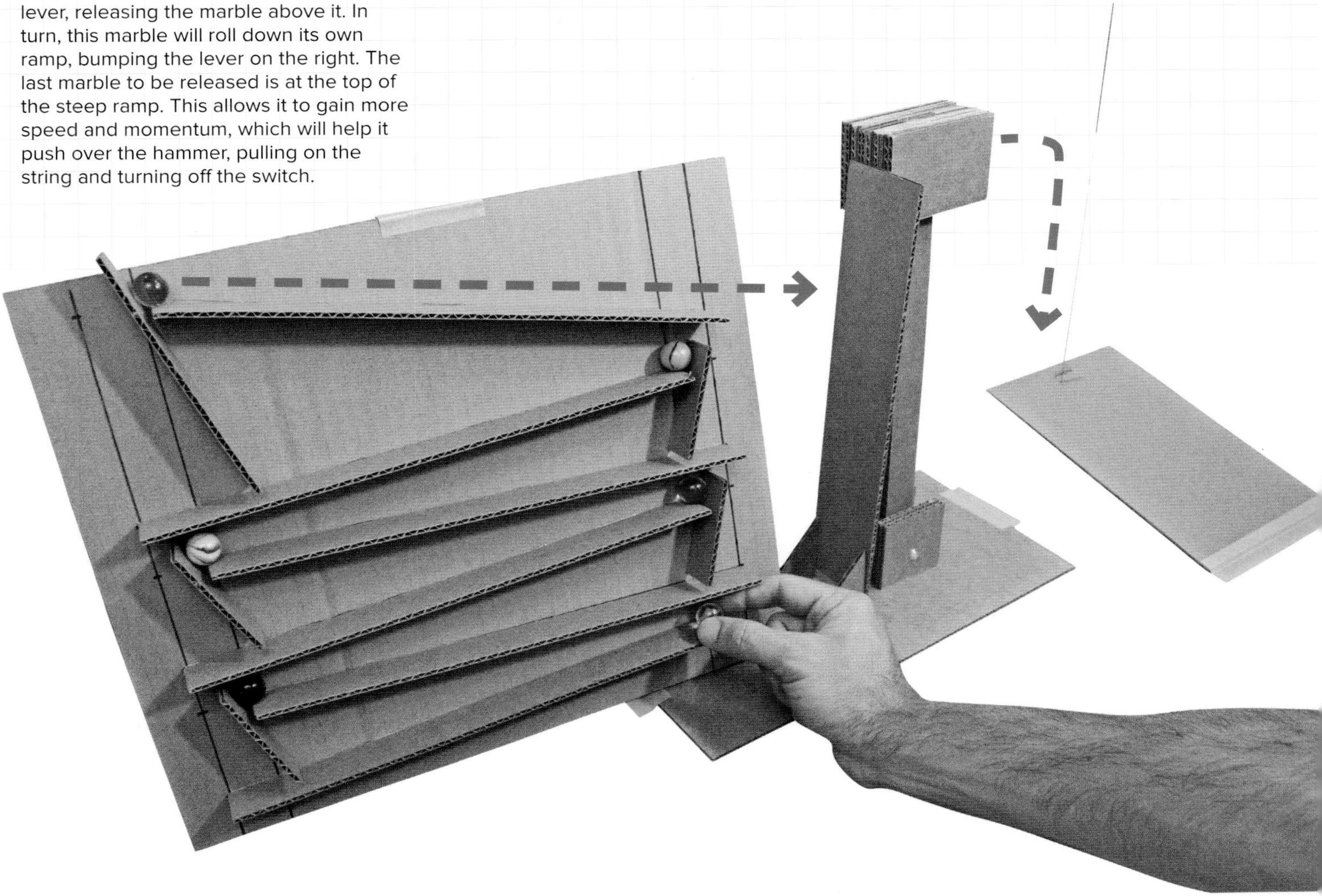

### Troubleshooting

If the marbles don't stay, you can push down or even cut out a small portion of the inside corner where the marble rests. Just make sure you don't remove too much, which might cause the marble to get stuck there.

### Engineering Know-How

The zig-zag ramp is a perfect example of a mini chain reaction. Once the first marble is set into motion, it triggers the next one. Because each marble is balanced at the top of the ramp, it only takes a little bump to cause them to roll down the ramp. And since the marbles are all the same size, the energy they pick up from rolling down the ramp creates enough of an extra force to bump the next marble, even though they have the same mass.

# DOOR OPENER

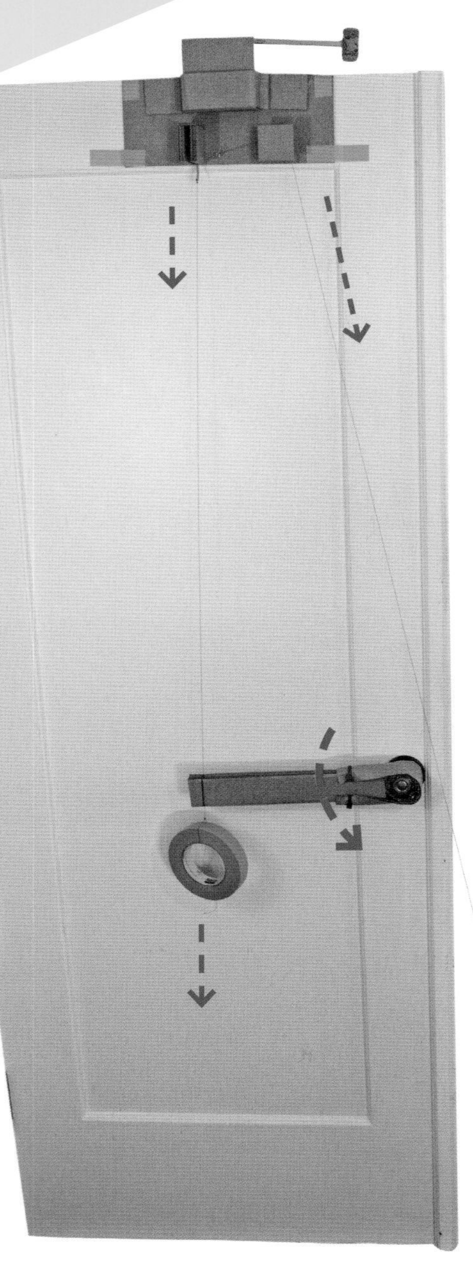

Can't use your hands because you're too busy juggling? Or maybe you forgot you filled your pockets with super glue and got stuck? No worries! Just make sure you build the Door Opener machine before you get into these sorts of situations.

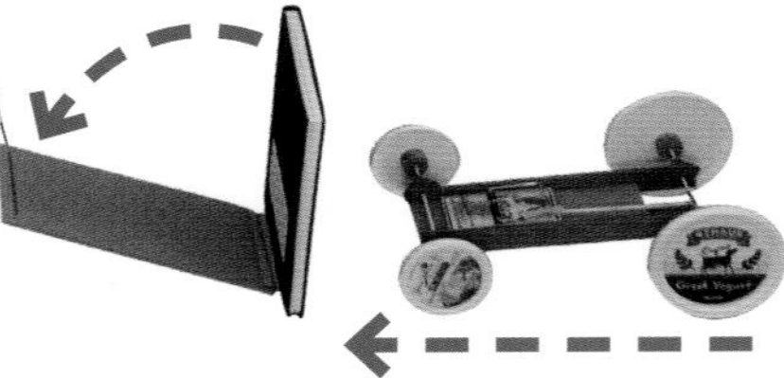

# BUILD MATERIALS & TOOLS

## DOOR-OPENING PIECES

**CARDBOARD:**

Six 2" x 12" (2.5 x 30.5 cm)–**A1, A2, A3, A4, A5 & A6**

Six 2" x 2" (5 x 5 cm)–**B1, B2, B3, B4, B5 & B6**

Six 2" x 4" (5 x 10.2 cm)–**C1, C2, C3, C4, C5 & C6**

One 1" x 2" (2.5 x 5 cm)–**D**

One 1" x 3" (2.5 x 7.6 cm)–**E**

Two 4" x 4" (10.2 x 10.2 cm)–**F1 & F2**

One 4" x 10" (10.2 x 25.4 cm)–**G**

One 5" x 12" (12.7 x 30.5 cm)–**H**

**OTHER:**

Mousetrap (the old-fashioned kind with metal pieces, not fake plastic cheese)–**I**

Zip ties or wire–**J**

Sharpened pencil–**K**

18' (5.5 m) strong string or fishing line–**L**

Cork (you can use cardboard if you don't have a cork)–**M**

Rubber bands or plastic wrap (optional)–**N**

Book

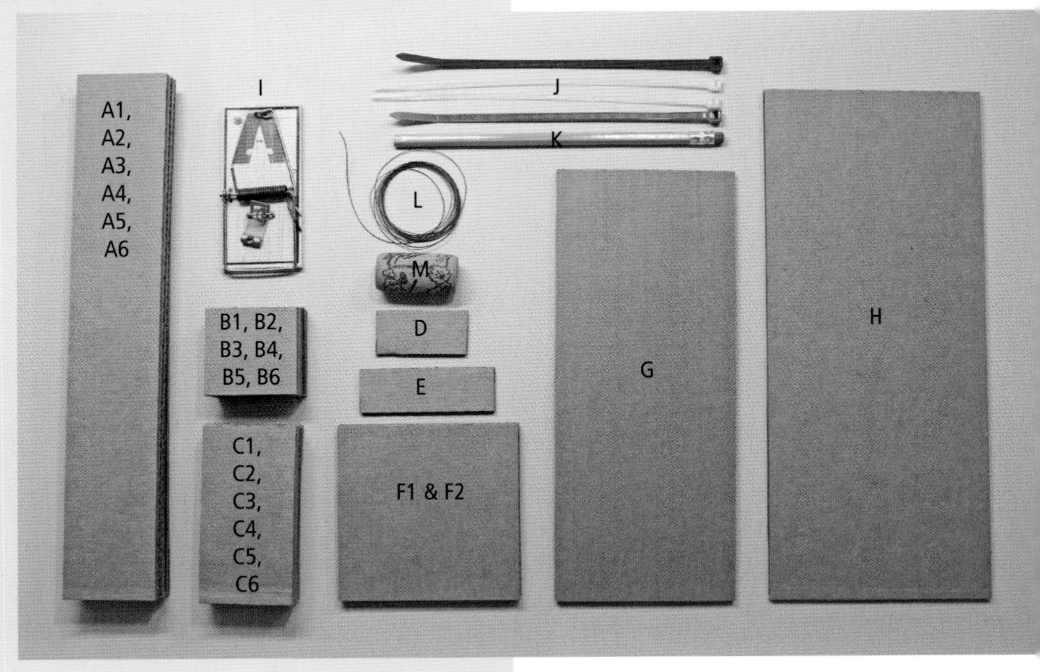


## MOUSETRAP-POWERED CAR

**CARDBOARD:**

Two 2" x 8" (5 x 20.2 cm) **O1 & O2**

Four 1" x 1" (2.5 x 2.5 cm)–**P1, P2, P3 & P4**

Two 1" x 10" (2.5 x 25.4 cm)–**Q1 & Q2**

**OTHER:**

Cork–**R**

Mousetrap (the old-fashioned kind with metal pieces, not fake plastic cheese)–**S**

3 skewers (Bamboo works best because it's really strong.)–**T**

4 plastic lids for wheels (I used two large in the back and two smaller ones in the front.)–**U**

String

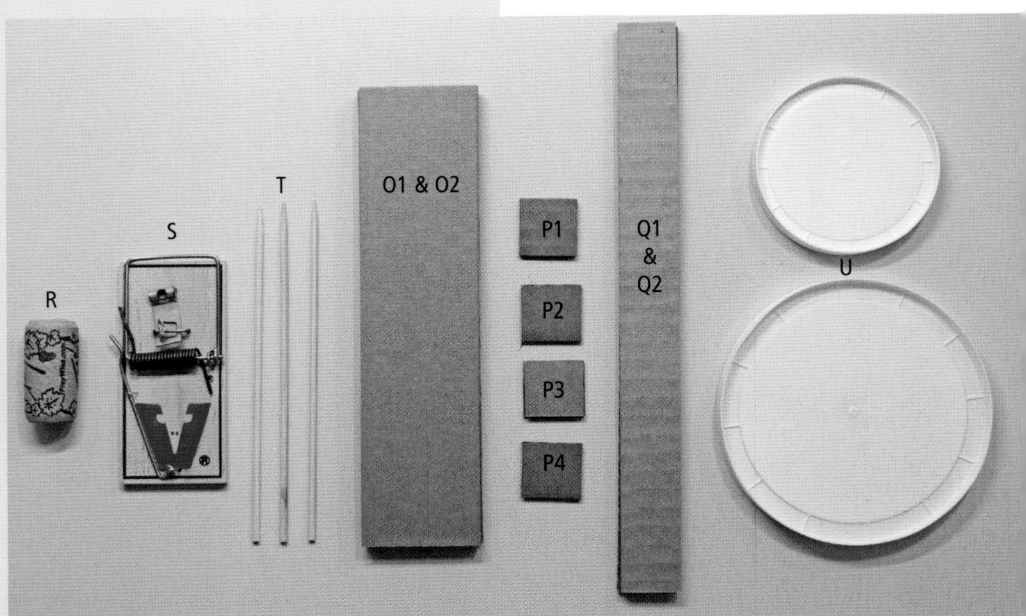


## TOOLS

Glue gun and hot glue

Pliers

Wire cutters

Awl

Painter's tape

Utility knife

Sharpened pencil or needle-nose pliers

Scissors

Roll of painter's tape

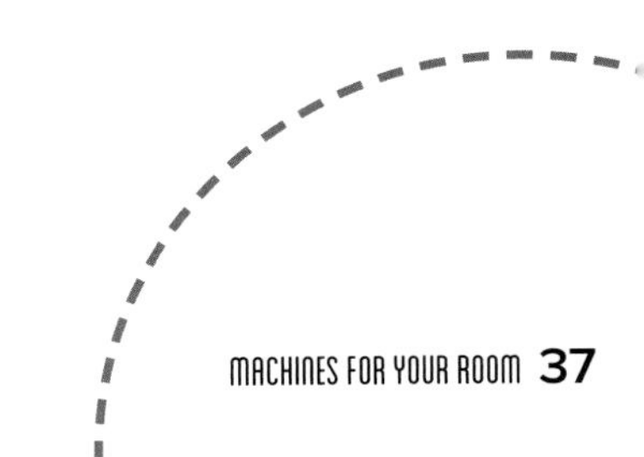

# BUILD THE MACHINE

### Make the Door Opener

1. Glue together C1, C2, C3, C4, C5, and C6. Glue together F1 and F2. Glue the stack of C pieces to the F pieces **(fig. a)**.
2. With pliers, remove the metal hold-down bar and the release pin from the mousetrap **(fig. b)**. Unfold and flatten the catch, and cut off the bent portion of the hold-down bar with wire cutters **(fig. c)**.
3. With an awl, poke a hole in the side of the cork and insert a sharpened pencil **(fig. d)**. Use hot glue to hold it in place.
4. Attach the pencil to the mousetrap hammer with zip ties or wire. Make sure it's oriented as shown. Glue the mousetrap to the C pieces as shown. Again, make sure you have the orientation correct. This is the portion that will push open the door **(fig. e)**.
5. Attach the door-opening mechanism to H **(fig. f)**.
6. For extra support, glue A1 over the stack of cardboard pieces **(fig. g)**.

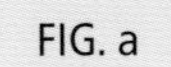
FIG. a

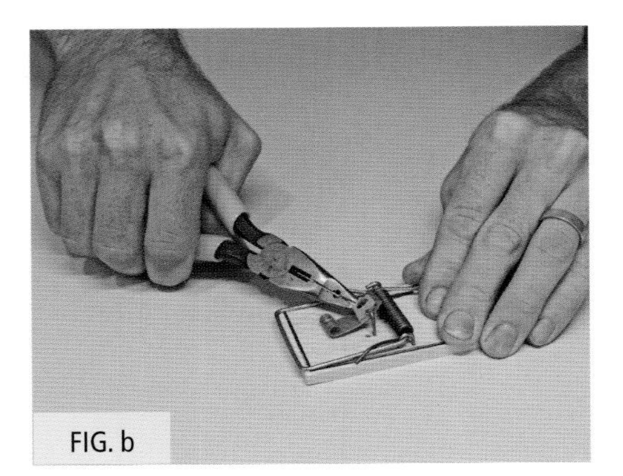
FIG. b

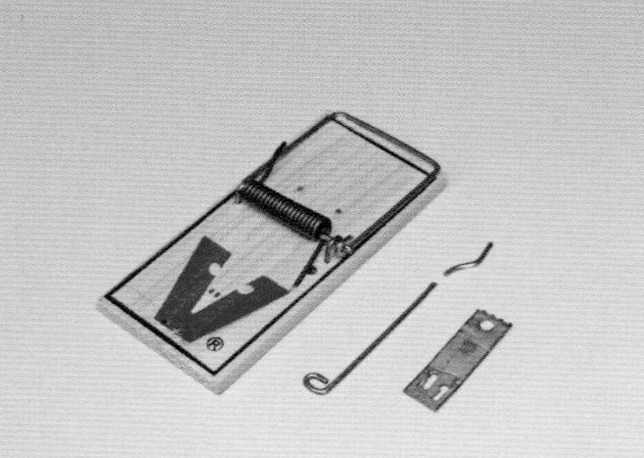
FIG. c

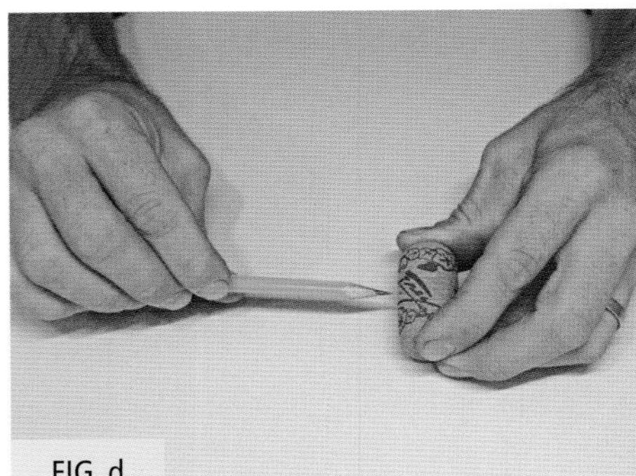
FIG. d

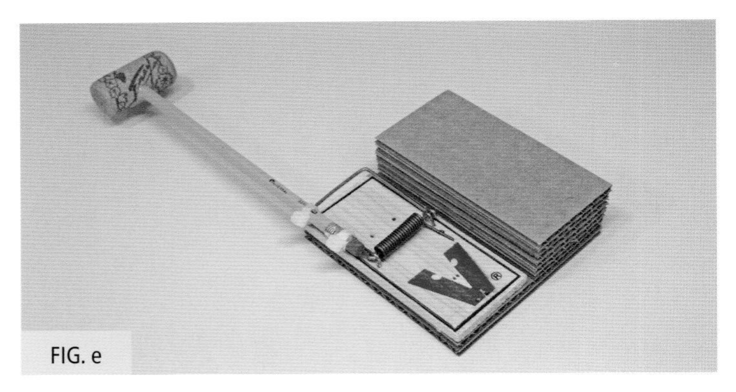
FIG. e

FIG. f

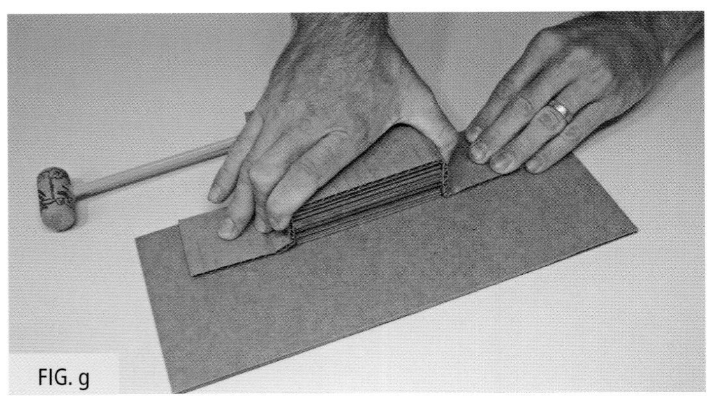
FIG. g

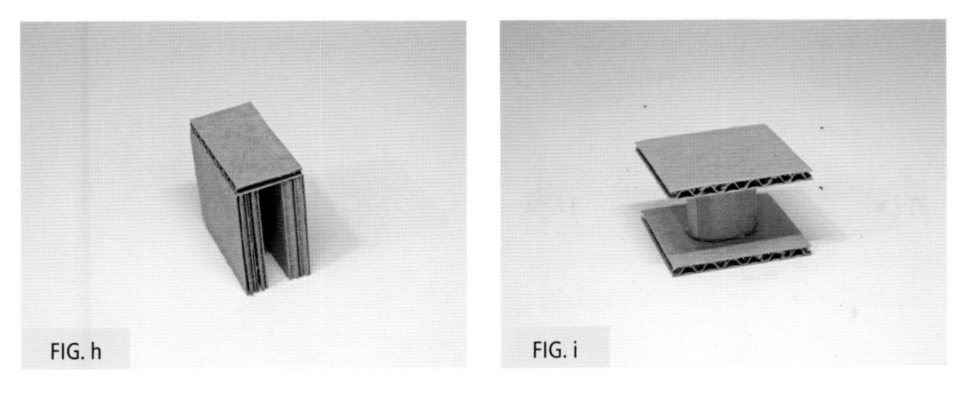
FIG. h

FIG. i

7. Divide B1, B2, B3, and B4 into two piles and glue each set together. Glue D to the edges of the B pieces, leaving a gap between the two doubled-up pieces **(fig. h)**. This will be where you attach your release pin.

8. Make a loop with E, secure it with tape, and glue that in between B5 and B6. This will be your string guide **(fig. i)**.

9. Glue the release pin and the string guide to H as shown in **fig. j**. You don't need to be exact with your placement; just use the bottom edge as a rough guide. Using an awl, poke a hole near the bottom of the pin release structure (glued B piles), approximately ½" x ½" (1.3 x 1.3 cm) from the corner, going all the way through to the other piece **(fig. j)**.

10. Glue together A2, A3, A4, and A5 in a stack **(fig. k)**. This will be the handle that attaches to your doorknob.

11. Poke three holes and cut three slits into G **(fig. l)**. The slits are so you can wrap and adjust your string without having to tie and cut knots. This is the platform that will ultimately pull the pin.

FIG. j

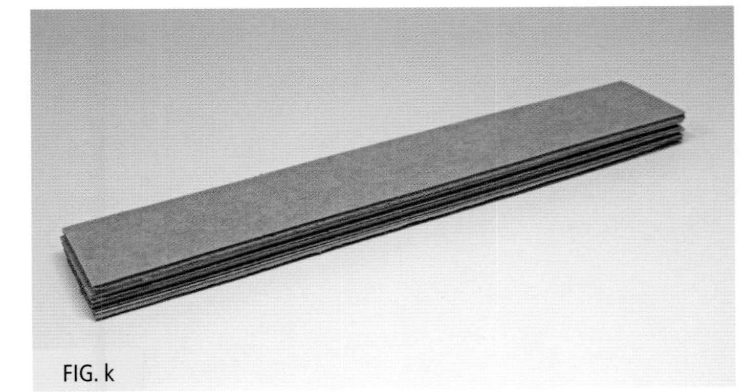
FIG. k

FIG. l

### Make the Doorknob Lever

1. Wrap A6 around your doorknob and attach it to the other glued A pieces using zip ties. Make sure it's as snug as you can get it. If it's too loose, it will spin around the knob and not actually turn it **(fig. m)**.

2. If you're having issues, you can wrap your doorknob in rubber bands or plastic wrap to give it some grip **(fig. n)**.

### Build the Mousetrap Car

1. Glue together O1 and O2 to make the base of the car. Glue Q1 and Q2 to the sides of the base **(fig. o)**.

2. Cut the cork into four equal pieces with a utility knife. Try to make the cuts as parallel to each other as you can. This will keep the wobble out of your wheels **(fig. p)**.

3. Poke holes into the sides of your car base. You want the skewers to rotate freely in the holes, so use a sharpened pencil or the tip of needle-nose pliers to gently enlarge each hole's diameter **(fig. q)**.

4. Most lids have a little bump exactly in the center. If your lid doesn't have a bump, or you're using another round object and don't know the center, use the center finder (see page 12). If your lid does have a bump, use an awl to poke a hole in the center. Make it large enough for the tip of your skewer to go through **(fig. r)**.

5. Poke a hole in the center of the cork. It doesn't need to be exactly in the center. The most important thing is that you keep the hole perpendicular to the cut you made. If you poke your hole at an angle, you're going to get wobble. Keep your awl poked through the cork as you glue it to the lid. This will keep everything lined up correctly **(fig. s)**. Do this for all four of your wheels.

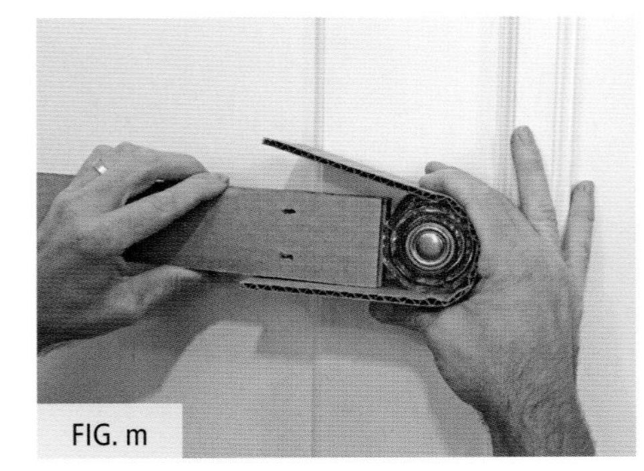
FIG. m

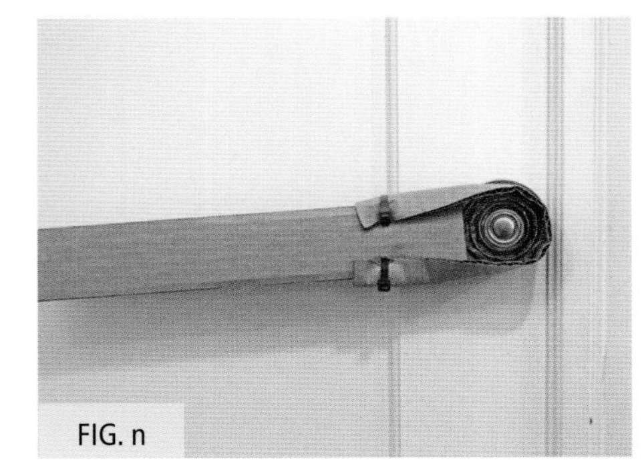
FIG. n

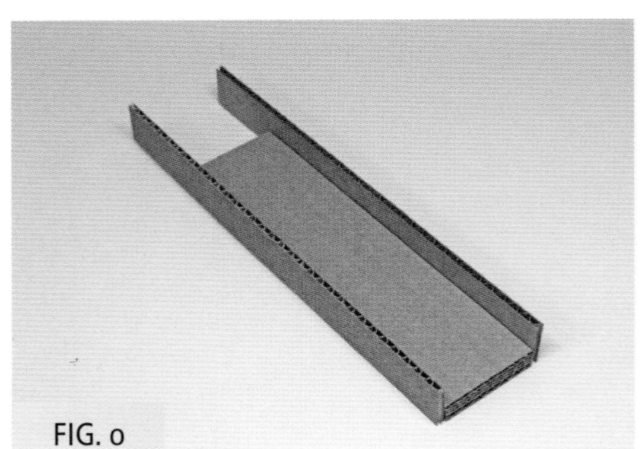
FIG. o

FIG. p

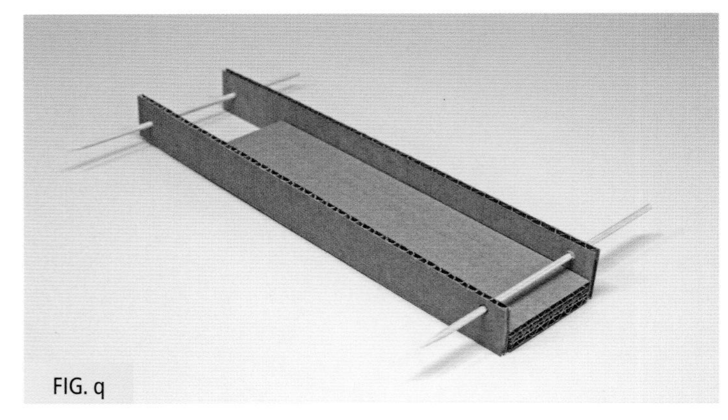
FIG. q

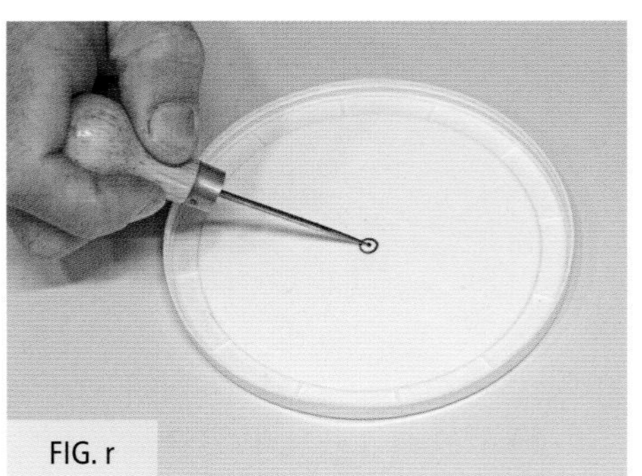
FIG. r

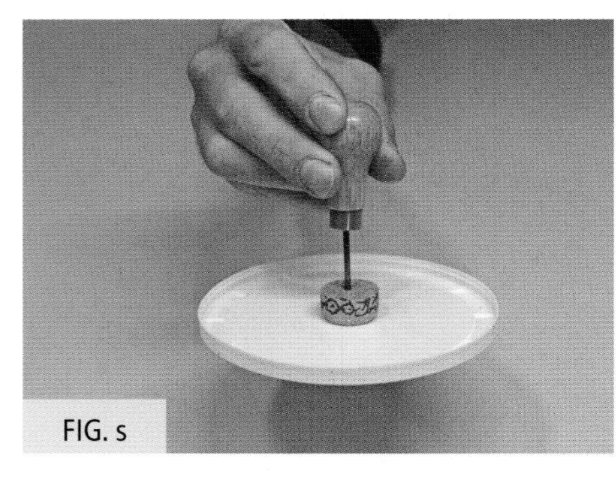
FIG. s

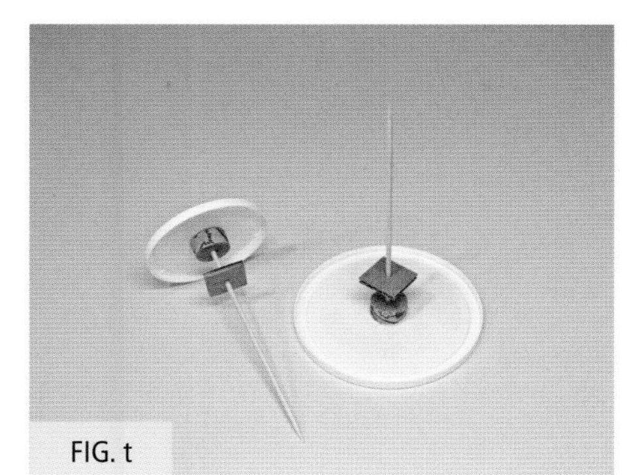
FIG. t

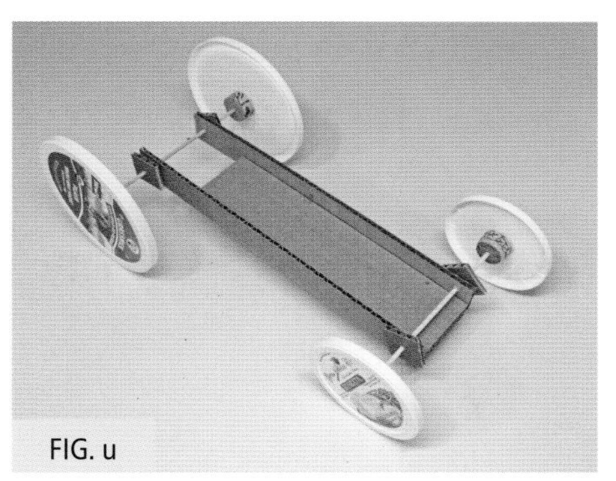
FIG. u

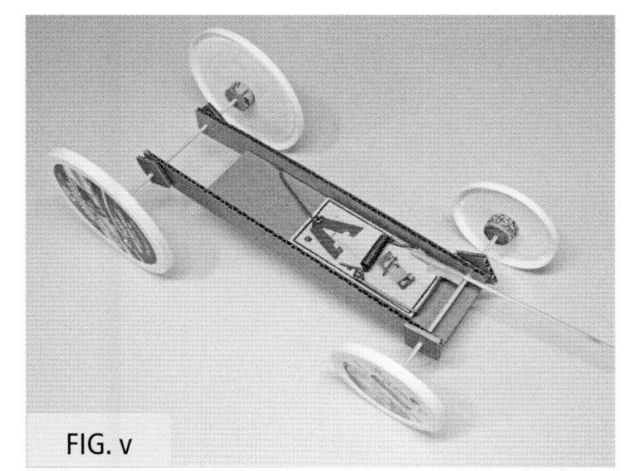
FIG. v

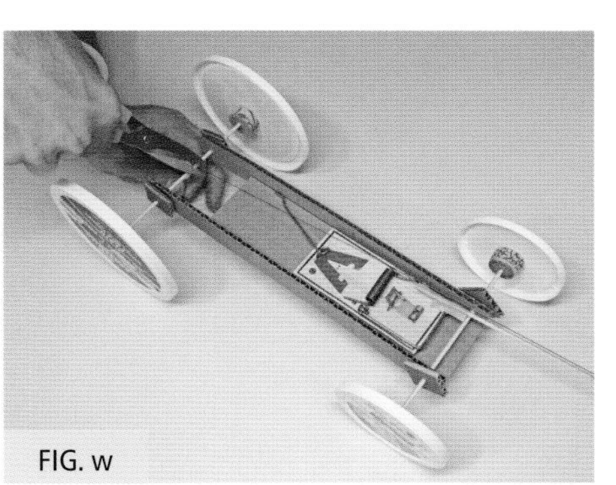
FIG. w

7. Insert one end of a skewer into one of the wheels. If one of the skewer ends is dull, you can use a pencil sharpener or utility knife to make it sharp. Poke holes into P1, P2, P3, and P4 and slide over the ends of the skewers **(fig. t)**. These will keep the axles lined up while the wheels spin.

8. Slide the ends of the skewers through the base of the car. Then slide on the other two spacers, followed by the wheels. The friction between the cork and skewers should be great enough to keep the wheels attached, but if it's not you can add a little glue. Adjust the spacers so the wheels are symmetrical **(fig. u)**.

9. Tape the last skewer to the mousetrap hammer. Make sure it's secure because this is the engine for your car. You can wrap wire or use zip ties as well, but make sure these don't get in the way of the hold-down bar **(fig. v)**.

10. Tie a piece of string to the end of the skewer. Add a drop of hot glue to make sure the knot doesn't come undone. Cut the string so it's just slightly short of the rear axle **(fig. w)**.

11. To wind the car, gently lift up on the skewer so the string can reach the rear axle. Twist the string around the axle until it rolls over itself, then keep rolling until the mousetrap is fully loaded **(figs. x and y)**. If you're having trouble getting your string to wrap around the skewer, add a little line of hot glue to the skewer to give it some grip. (Don't glue the string to the skewer!)

12. Cut the remaining string in half and tie each one to the mousetrap pieces as shown in **fig. z**. These pieces are what will release a weight that will turn the doorknob.

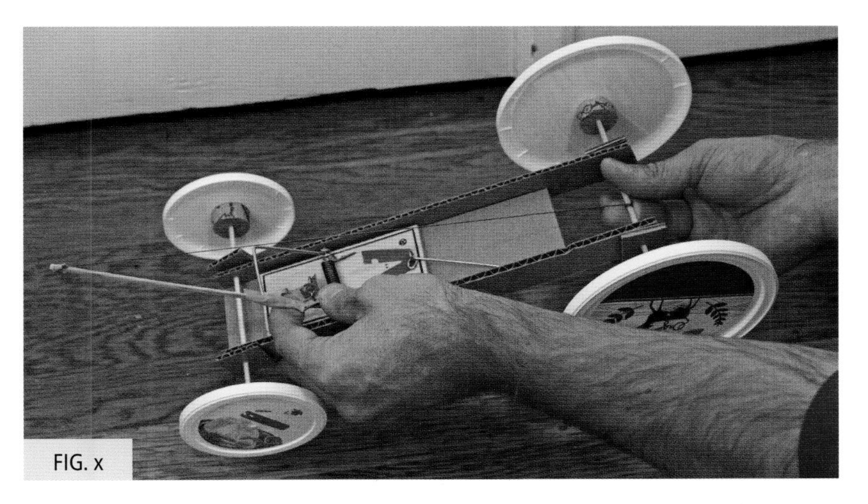
FIG. x

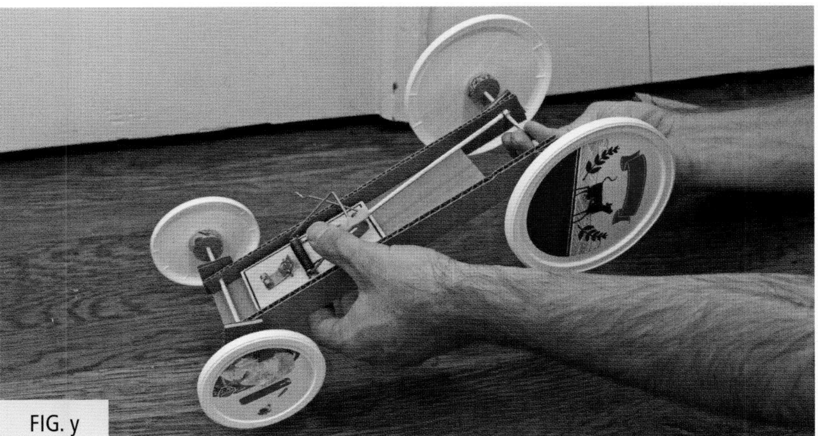
FIG. y

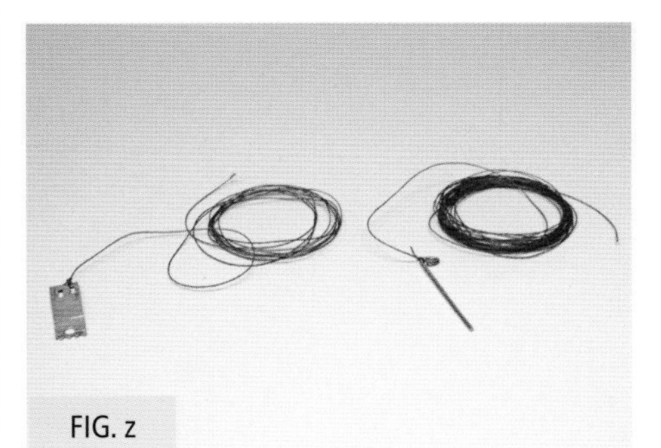
FIG. z

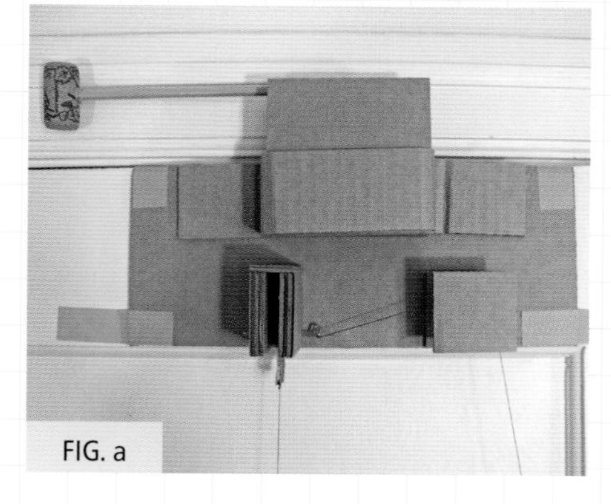

FIG. a

# START THE CHAIN REACTION!

### The Setup

1. Tape the door opener mechanism to the top of your door **(fig. a)**. Slide the flat metal piece (the one you removed from the mousetrap and flattened) between the gap in the B pieces and slide in the pin from the right side.

2. Tie the string from the flat piece to the doorknob lever. Make sure the lever is horizontal and the string is fairly taut. Attach a weight to the end of the lever. I used a roll of tape **(figs. b and c)**. Use something heavy enough to pull down the lever, which turns the knob. Don't make it too heavy or it will either break the lever or cause it to slide off the doorknob.

FIG. b

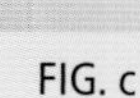

FIG. c

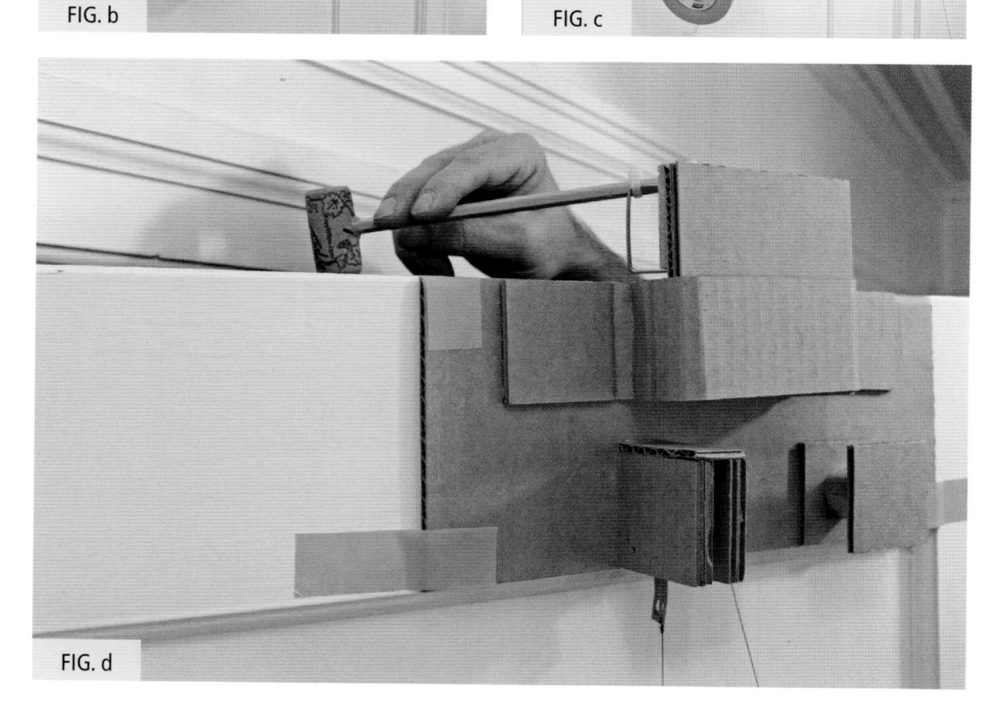

FIG. d

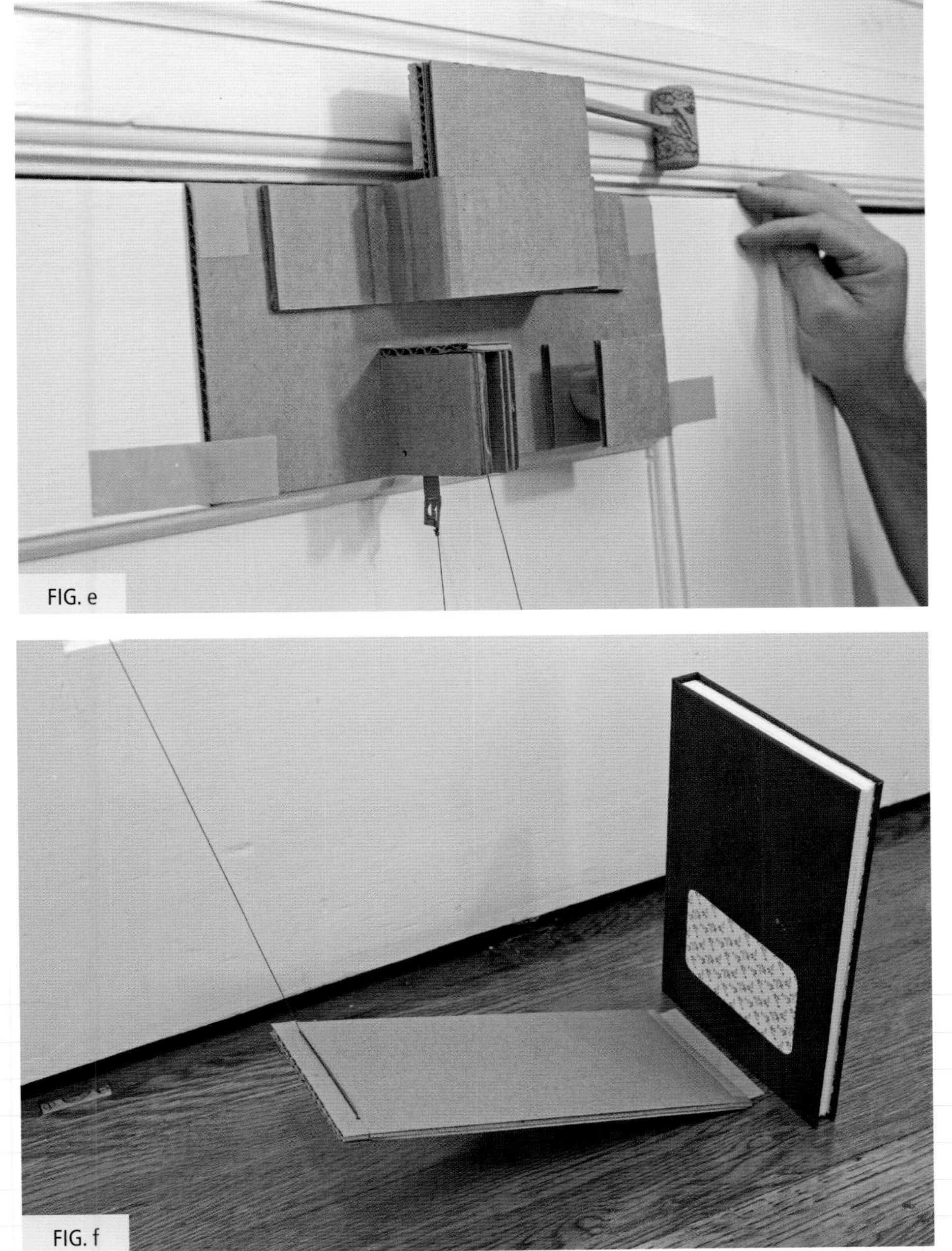
FIG. e

FIG. f

3. The spring from the mousetrap, along with the pencil, is what will open your door once the knob is turned. To set the spring, simply open the door wide enough to pull the pencil to the other side **(fig. d)**. You'll probably need to stand on a stool or chair to reach. When the pencil is on the other side, close the door so it latches **(fig. e)**.

4. Tape G to the floor to create a lever. Attach the pin side of the string to the end of this piece. You can wrap the string around and through the slits you cut so you can adjust it without having to cut your string. You don't need the angle to be too steep, just enough to pull out the pin when it's flattened by the book. Place a book at the base of the lever **(fig. f)**.

**Make It Go!**

To set off the chain reaction, set up your mousetrap car so it'll drive into the book, causing the book to fall over. The weight of the book flattens the lever, pulling out the pin. This releases the weighted lever, which turns the doorknob. Once the door latch is disengaged, the loaded spring on the top of the door starts pushing it open.

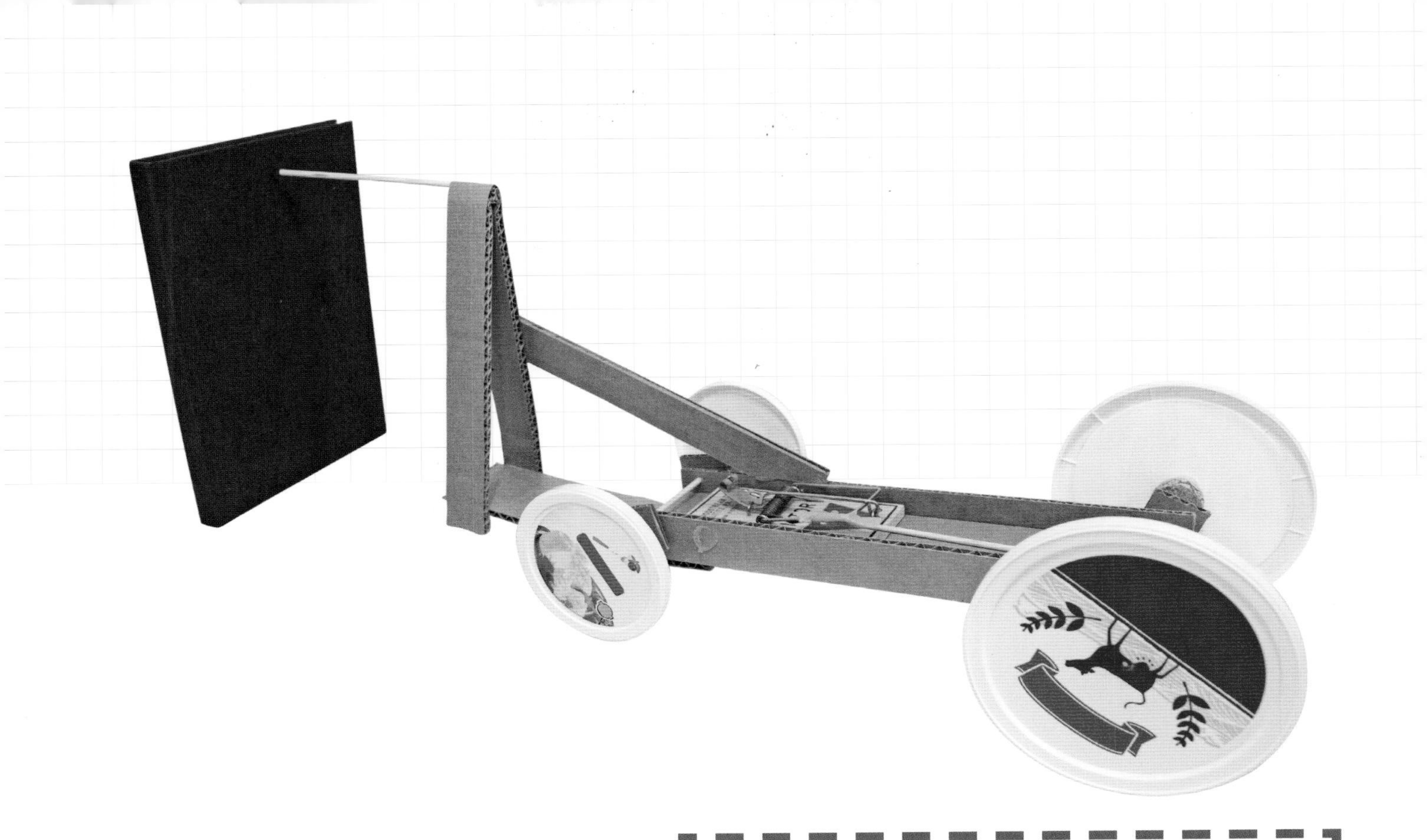

## Engineering Know-How

The Door Opener employs many examples of potential energy. When you set the mousetrap on the car, you're turning the wound spring into potential energy. There's also a mousetrap in the door opening mechanism that works the same way. The book (in its upright position) and the weight that's attached to the door handle are also potential energy. Nothing happens when they're left alone, but the second something interferes (the book being bumped or a pin being pulled) gravity takes over and turns the mass of the objects into a force.

## Troubleshooting

If your mousetrap car can't push over your book, you can attach a skewer to the front to act as a mini battering ram. This will hit the book above its center of gravity, causing it to fall over more easily. You can also move the door-opener mechanism to alter how fast and far the door will open. If you move it to the hinged side of the door, it will open farther but at a much slower rate, if at all. Older doors, like mine, are very heavy because they're made of solid wood. Mine also has a mirror attached to it, which makes it even heavier. If you move the door-opener mechanism to the other side, it'll be easier to open but will only go as far as the length of the pencil. If you make the pencil longer, it might not have enough force to open it at all. To add more force to the opener, you could use a rat trap, which is larger, or use two mousetraps in tandem.

# CHAPTER III

# MACHINES FOR AROUND THE HOUSE

While not as convenient as your own personal robot, these machines take the monotony out of simple chores that you do every day. You'll never see your toothbrush as boring and ordinary again!

The Toothpaste Squeezer (see page 58) makes brushing more exciting.

# PLANT WATERER

So you forgot to water the plants and now they're all dead. I get it. These things happen. Until now! Build this plant-watering machine and quench your plants' thirst. Just make sure you don't forgot to set off the chain reaction to get the machine going.

# BUILD MATERIALS & TOOLS

### MARBLE RAMP

**CARDBOARD:**
One 2" x 6½" (5 x 16.5 cm)–**A**
One 2" x 2" (5 x 5 cm)–**B**
One 2" x 6" (5 x 15.2 cm)–**C**
One 2" x 12" (5 x 30.5 cm)–**D**

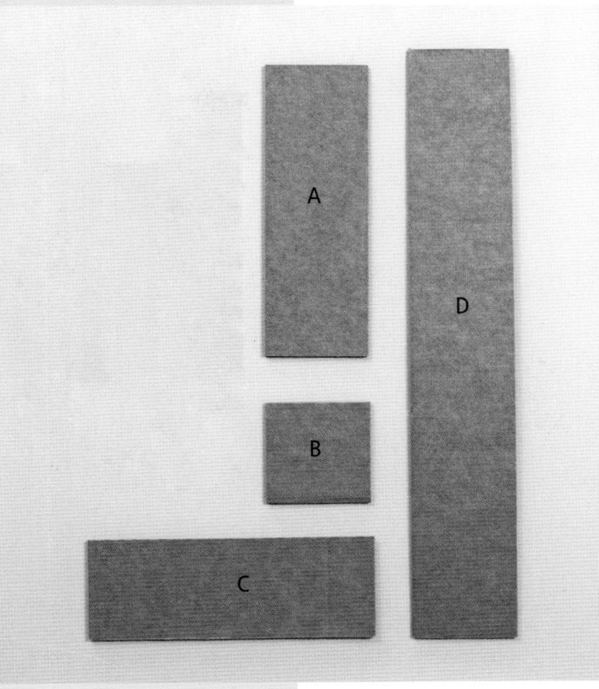


### BALLOON LOWERER

**CARDBOARD:**
One 1" x 5" (2.5 x 12.7 cm)–**E**
Four 1" x 1" (2.5 x 2.5 cm)–**F1, F2, F3 & F4**
Two 2" x 2" (5 x 5 cm)–**G1 & G2**
Three 1" x 2" (2.5 x 5 cm)–**H1, H2 & H3**
Three 1" x 4" (2.5 x 10.2 cm)–**I1, I2 & I3**
One 2" x 5" (5 x 12.7 cm) thin cardboard–**J**
One 4" x 9" (10.2 x 22.9 cm)–**K**

**OTHER:**
Two 6" (15.2 cm) skewers–**L**
Marble–**M**
Balloon–**N**

*(continued)*

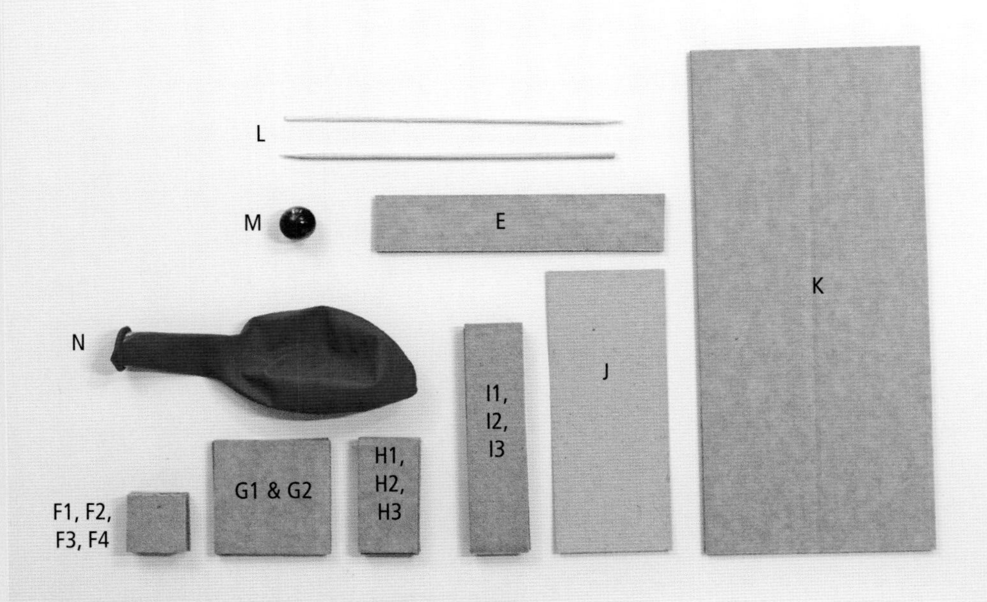


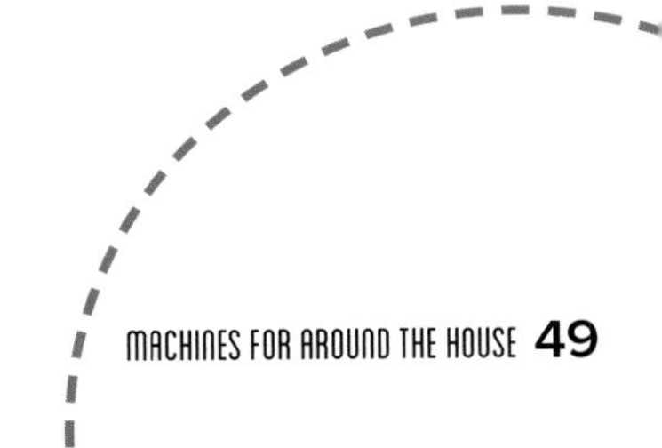

## WATERING MECHANISMS

### CARDBOARD:

Two 5" x 13" (12.7 x 33 cm)–**O1 & O2**

Six 2" x 7½" (5 x 19 cm)–**P1, P2, P3, P4, P5 & P6**

Three 2" x 6" (5 x 15.2 cm)–**Q1, Q2 & Q3**

Four 1" x 1" (2.5 x 2.5 cm)–**R1, R2, R3 & R4**

One 4" x 4" (10.2 x 10.2 cm)–**S**

One 2" x 4" (5 x 10.2 cm), cut at an angle to make two triangles–**T**

### OTHER:

3' (1 m) string–**U**

2 pieces of 1½" (3.8 cm) wire, each bent into a U shape–**V1 & V2**

Four 6" (15.2 cm) skewers or two 12" (30.5 cm) skewers–**W**

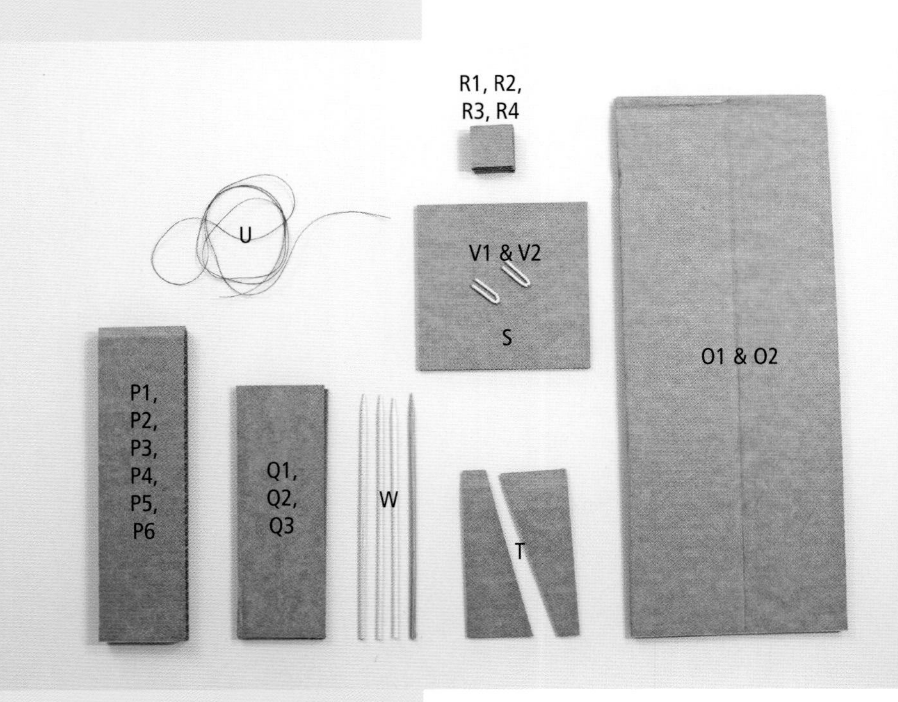


## PLANT AND WATERING CANS

### CARDBOARD:

One 8" x 12" (20.3 x 30.5 cm)–**X**

### OTHER:

Large plastic cup–**Y**

2 or 3 small plastic cups (The third one is optional, depending on the type of plant you have.)–**Z**

Plant to water–**AA**

## TOOLS

Marker

Ruler

Scissors

Utility knife

Glue gun and hot glue

Awl

Sharpened pencil

Painter's tape

Pencil or small screwdriver

# BUILD THE MACHINE

### Make the Marble Ramp

1. Measure and cut a ½" x 1¼" (1.3 x 3.1 cm) section from the tops of C and A **(figs. a and b)**.
2. With a utility knife, score D on each side ½" (1.3 cm) from the edges (see page 14). Crease at the score lines to create a trough **(fig. c)**.
3. Glue A to the front of the trough, and glue C about 4" (10.2 cm) from the rear. Add B to A to give it more stability **(fig. d)**.

### Make the Balloon Lowerer

1. Glue H1 and H2 to the sides of I1 and cap the ends with F1 and F2 **(fig. e)**.
2. With an awl, poke holes in the center of G1 and G2 about ½" (1.3 cm) from the top edge. Expand the hole using a pencil, and make it large enough so two skewers can rotate inside of it. Glue them to the base (K), 1" (2.5 cm) from the edge and 4" (10.2 cm) apart **(fig. f)**.
3. Insert the skewers and tape the ends to keep them together. Fold E in half and glue it to the skewers at the location shown. Pay attention to the orientation. When the cardboard piece is all the way forward (like it is in the image), the skewers should be stacked on top of each other. Glue I2 and I3 together and then glue them inside of the folded piece **(fig. g)**.

FIG. a

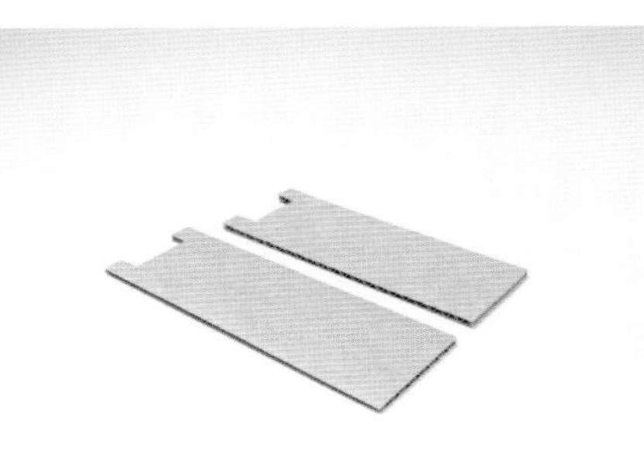

FIG. b

FIG. c

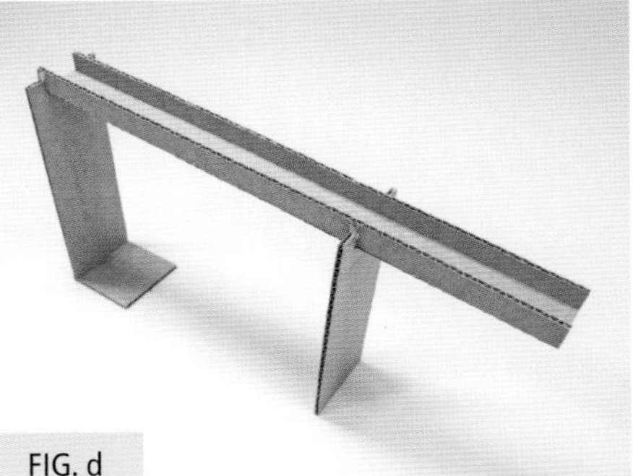

FIG. d

FIG. e

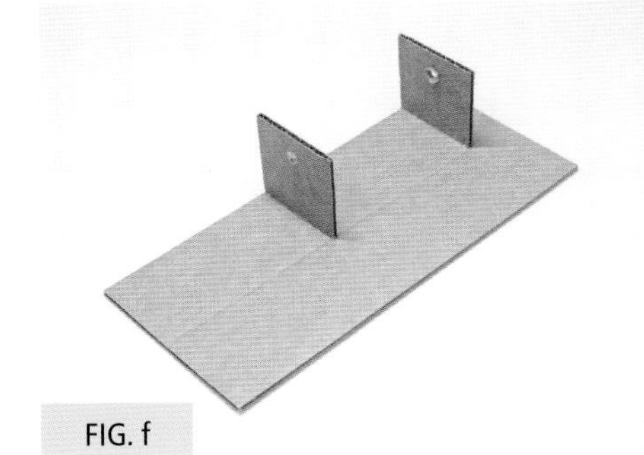

FIG. f

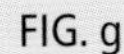

FIG. g

4. Attach the marble trough and add H3 and F3 and F4 for support **(figs. h and i)**.
5. Roll the thin piece of J into a tube and cover it with tape. Insert the balloon in between the skewers, paying attention to the orientation. You might need to pry apart the two skewers using a pencil or small screwdriver in order to fit the balloon. Once it's in place, center it between the ends of the skewers and insert the tube **(fig. j)**.
6. To make sure the balloon leaks air correctly, blow into the tube to inflate it (the marble trough lever mechanism should be lowered so the skewers are stacked one above the other). When you stop blowing into the balloon, it should slowly deflate. If it feels impossible to deflate, or deflates extremely slowly, you can add little spacers, using a thin piece of cardboard **(fig. k)**.
7. Inflate the balloon again, and this time, lift up on the trough lever. Now the skewers should be side by side. This effectively cuts off the flow of the balloon by pinching it at the inlet **(fig. l)**.

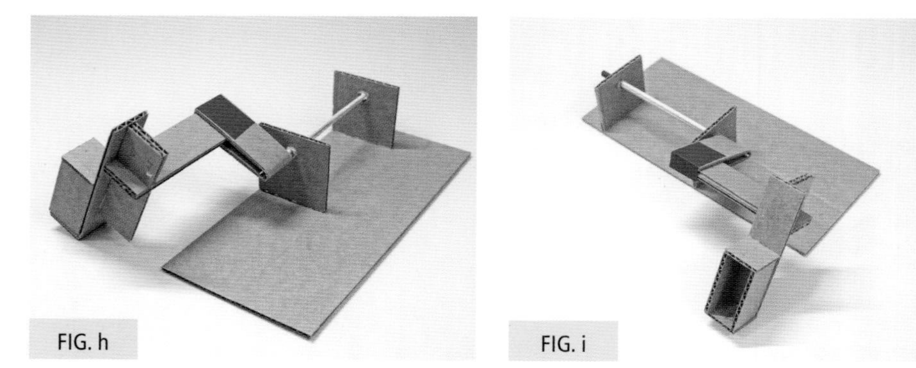
FIG. h

FIG. i

FIG. j

FIG. k

FIG. l

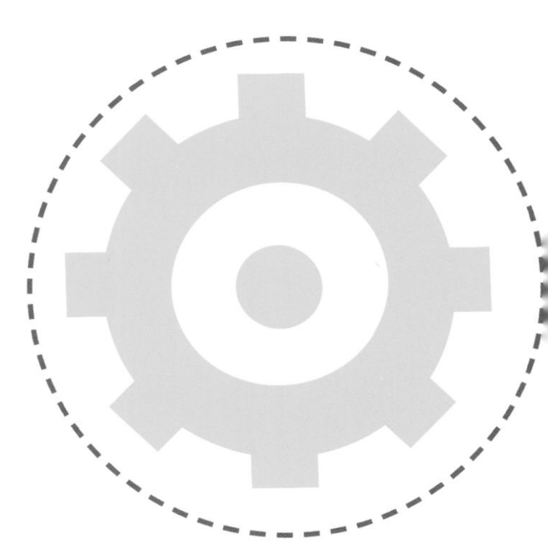

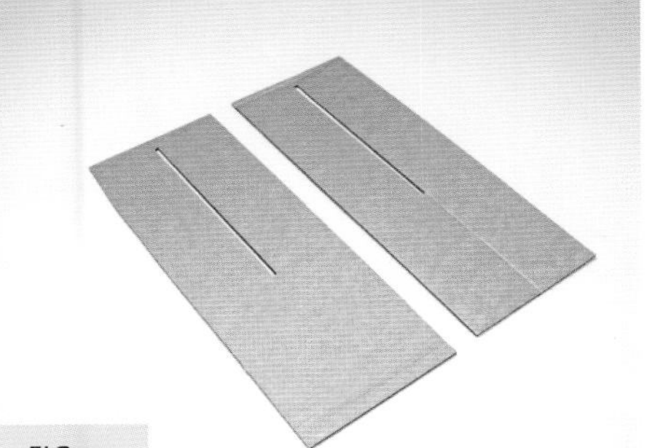
FIG. m

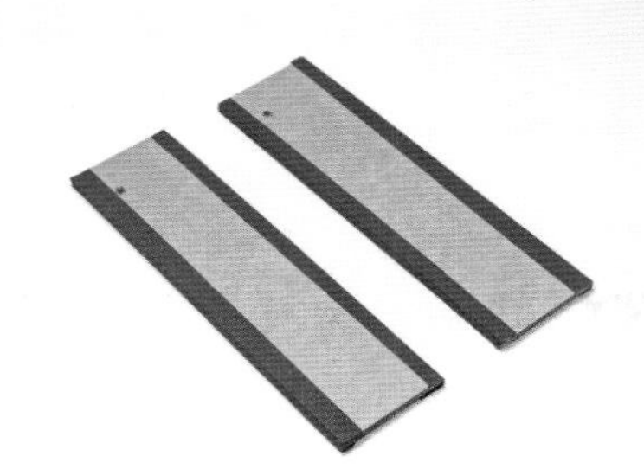
FIG. n

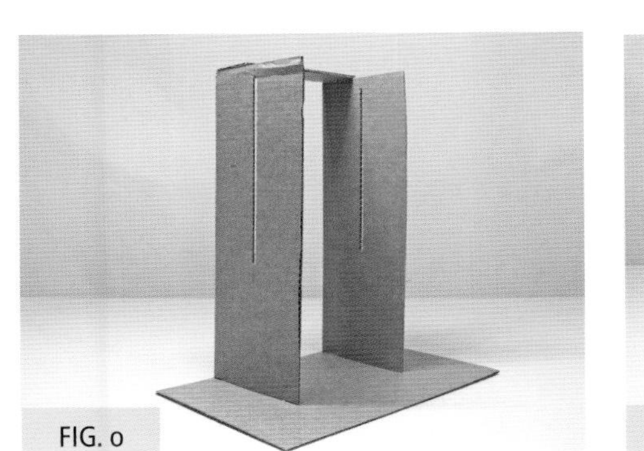
FIG. o

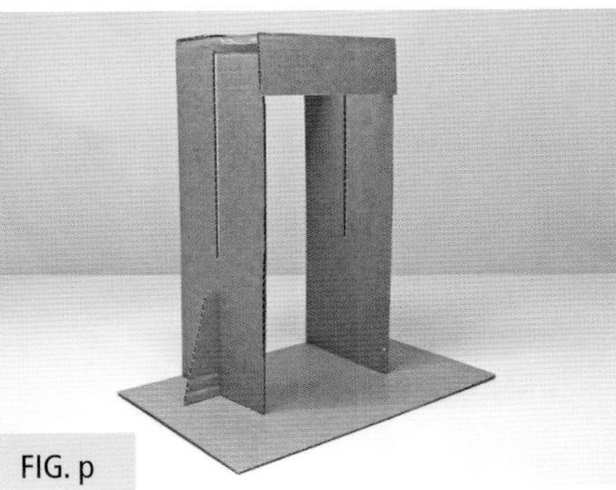
FIG. p

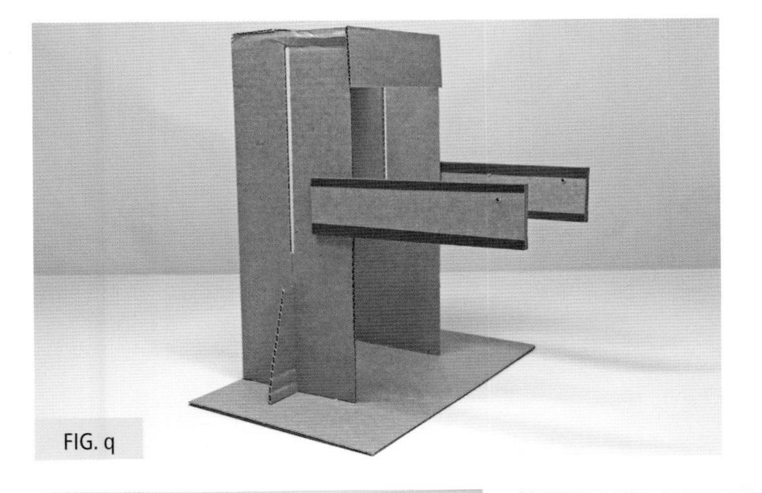
FIG. q

FIG. r

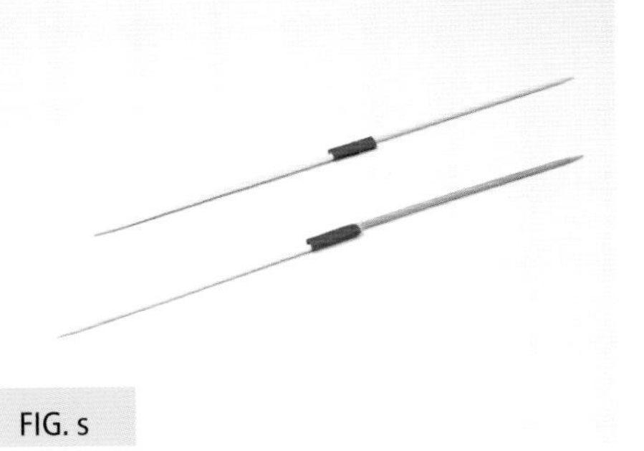
FIG. s

### Build the Main Structure

1. Cut a slit into O1 and O2. Start 1" (2.5 cm) from the top and make it 7" (17.8 cm) long. Make sure it's wide enough for a skewer to slide up and down easily **(fig. m)**.

2. Take P1, P2, P3, and P4, and tape them together to create two pairs, each two layers thick. Poke a hole 1" (2.5 cm) from the short edge and ¾" (1.8 cm) from the bottom **(fig. n)**.

3. Glue the uprights to the base (X), centered along one edge and 6" (15.2 cm) apart. Glue Q1 support piece to the top as shown in **fig. o**.

4. Add the Q2 support to the front edge and glue the triangular pieces (T) to the bottom **(fig. p)**.

5. Glue on the thick cardboard arms (made from the P pieces) near the bottom edge of the slit. Make sure you leave a ¾" (1.8 cm) gap between the slit and the edge of the arm **(fig. q)**.

6. Cut a hole in S that's slightly smaller than the largest part of the cup. Also make sure it's close to one edge of S. This is where we'll attach a skewer so it can pivot **(fig. r)**.

7. Glue and tape together two pairs of the 6" (15.2 cm) skewers (or use two 12" [30.5 cm] skewers) **(fig. s)**.

8. Center one of the skewer pieces on the side of S nearest the cutout and attach with glue and tape. Poke holes in your larger container and insert the other skewer **(fig. t)**.
9. Insert the skewer attached to the cup to the front arms. Place the larger container with the skewers into the slits of the uprights. Poke holes in the centers of R1, R2, R3, and R4 to make washers (see page 17) and slide them over the ends of all of the skewers, making sure they slide and rotate freely **(fig. u)**.
10. Add P5 and P6 to the top of the structure and place the Q3 support piece on the top **(fig. v)**.
11. Insert the U-shaped wires into the cardboard at the top as shown, and secure with glue, making sure you don't get glue on the part the string goes through. Cut out a section in the middle support so the string can move freely. Tie one end of the string to the skewer attached to the larger container, and then run it through the looped wires **(fig. w)**.

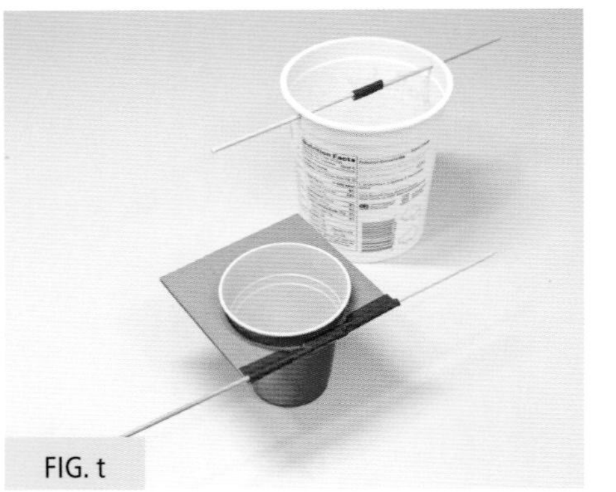
FIG. t

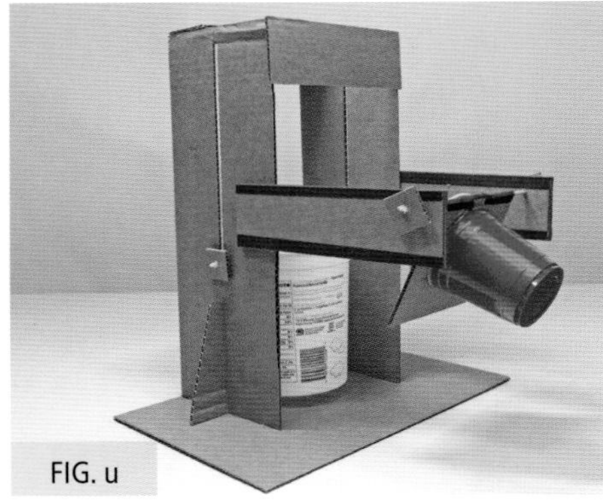
FIG. u

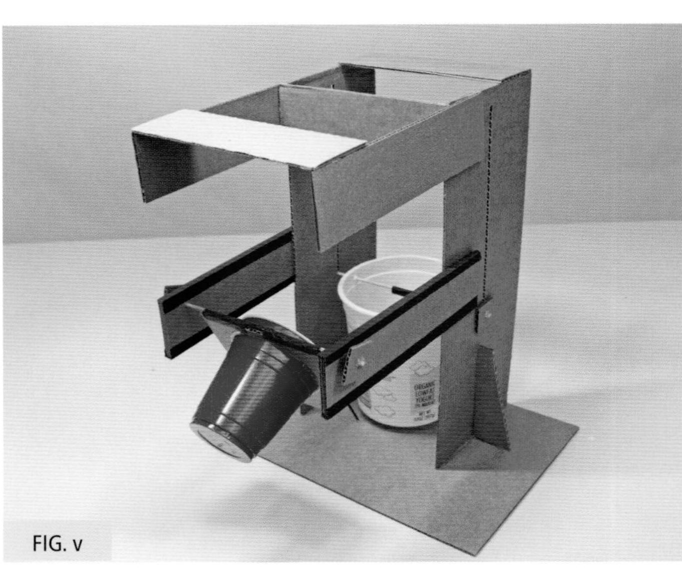
FIG. v

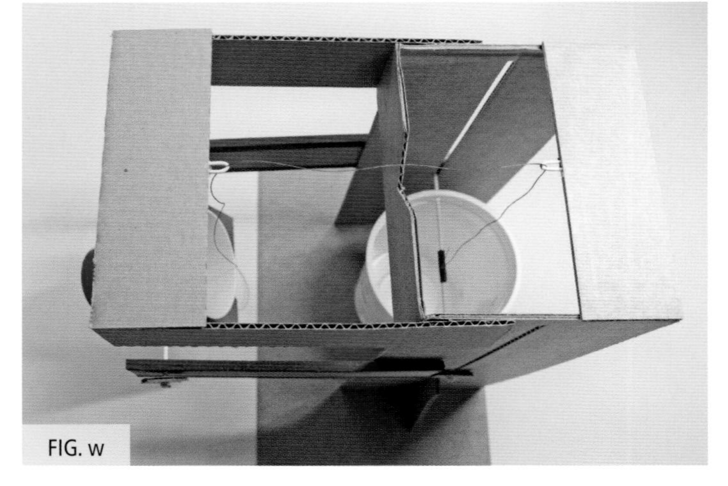
FIG. w

12. Lift up the larger container until it can go no higher, and then attach the other end of the string to the bottom of the small cup. Use glue and tape to make sure it's on there good and tight. You might need to cut a notch in S as well so the string can make a straight line to the bottom of the cup **(fig. x)**. This is easier to do if you flip the structure upside down **(fig. y)**.

13. When the larger container is pushed up all of the way, the smaller cup should be flat on top. As the container is lowered, the cup will start to tilt, and when the container is all the way at the bottom, the cup is fully tilted. Instead of lowering the container by hand, we'll add in the balloon, which will allow the container to slowly lower as the air leaks out of the balloon **(figs. z and aa)**.

FIG. x

FIG. y

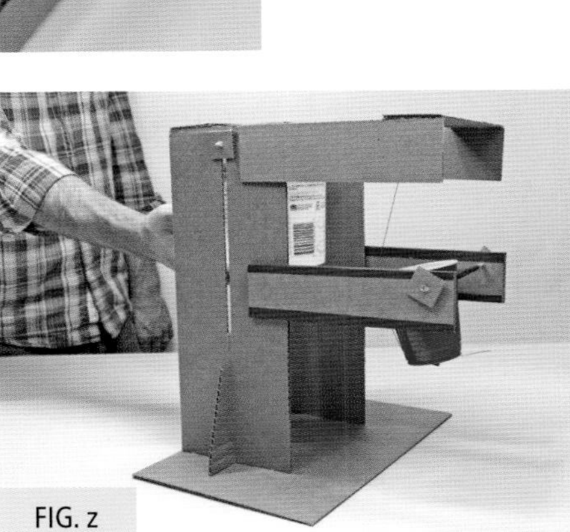

FIG. z

FIG. aa

FIG. a

# START THE CHAIN REACTION!

FIG. b

### The Setup

1. Inflate the balloon and place it under the large container. Make sure the trough lever is horizontal so it's cutting off the airflow of the balloon. Now line up the end of the marble ramp with the top of the trough **(fig. a)**.
2. Place the plant in front of the cup so that, when the cup is fully tipped, it will pour water into the plant **(fig. b)**.
3. Instead of trying to pour water directly into the attached cup, it's easier to pour water into a separate cup first, and then nest that inside of the other one **(fig. c)**.
4. Add a weight or even some water into the larger container. The trick it to make it heavier than the small cup filled with water.

FIG. c

## Make It Go!

With the balloon inflated and water in the cup, you can set off the chain reaction by placing a marble at the top of the ramp. When it rolls to the end, it will fall into the trough. The weight of the marble should be enough to cause the trough to fall over, which will allow air to slowly escape from the balloon. As the balloon deflates, the large container will gently be lowered. Because it's heavier than the cup, as it lowers it will tilt the cup. When the balloon is fully deflated and the container is fully lowered, the cup will have emptied all of its water into the plant.

## Troubleshooting

If you're having an issue with your trough lever staying horizontal, add a bit of friction by pushing a shim up against the edge of the cardboard and then covering it with tape. The trick is to find the balance that gives you enough friction to keep the trough lever in the right spot, but have it fall over when the marble is released.

Other issues you might have are things not running smoothly. This might mean your track is too narrow, or some of your holes aren't large enough to allow things to spin freely. It's an easy fix, but it might take some patience on your end to suss out the culprit.

## Engineering Know-How

Pulleys are typically used to change the force that's needed to lift an object, or they change the direction of that force. In this machine, the bent loops of wire act as pulleys that cause the falling yogurt container (downward force) to lift the end of the water cup (upward force).

# TOOTHPASTE SQUEEZER

Just because you brush your teeth twice a day doesn't mean you can't make it a little more exciting. This machine not only adds toothpaste to your bristles, but it also brings your toothbrush to the perfect toothbrush-grabbing position.

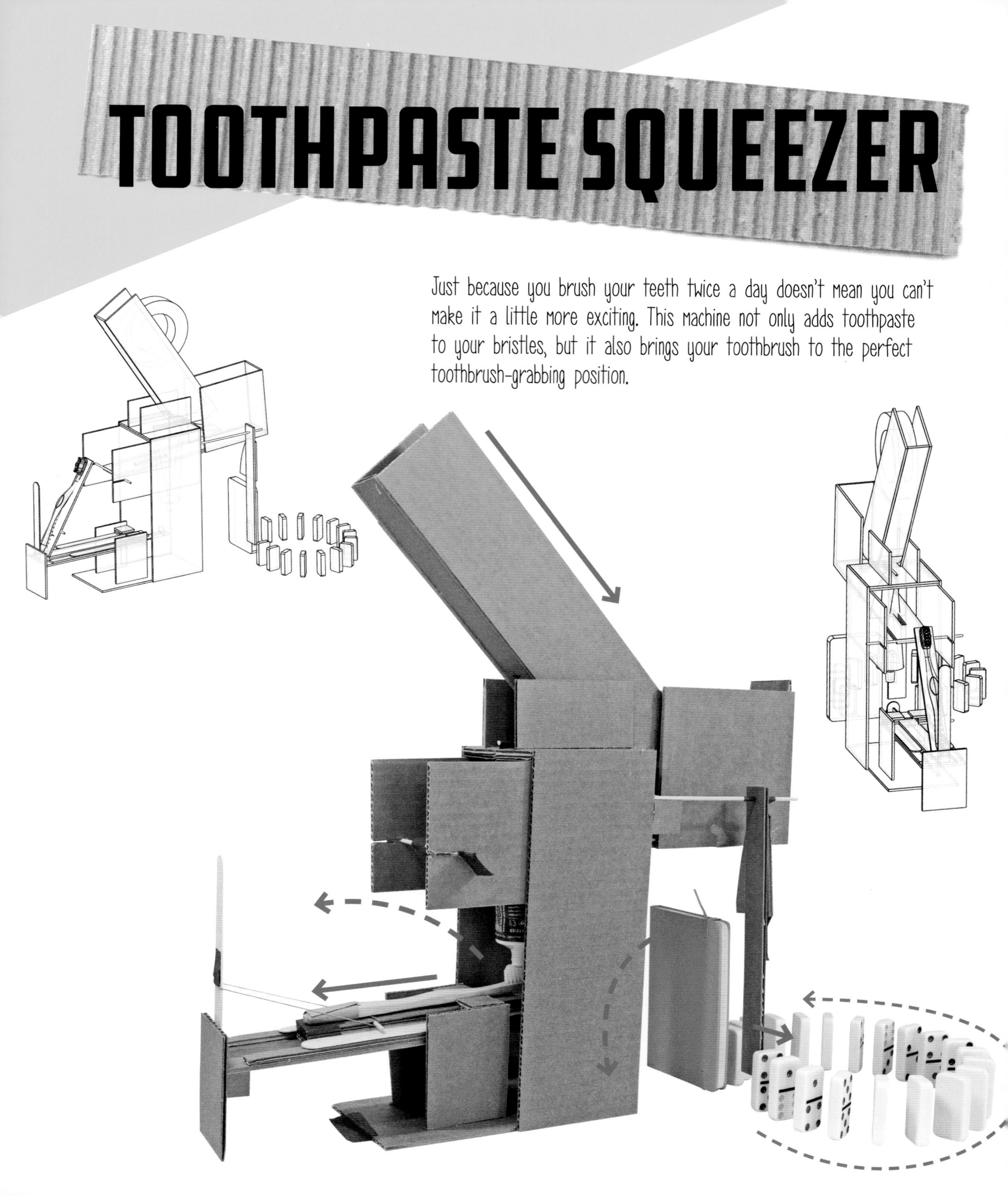

# BUILD MATERIALS & TOOLS

### TOOTHBRUSH SLIDER

**CARDBOARD:**

Two 1" x 1" (2.5 x 2.5 cm)–**A1 & A2**
Four ¾" x 1" (1.8 x 2.5 cm)–**B1, B2, B3 & B4**
Three 2½" x 3½" (6.4 x 8.9 cm)–**C1, C2 & C3**
Two ½" x 8" (1.3 x 20.3 cm)–**D1 & D2**
Two 1" x 8" (2.5 x 20.3 cm)–**E1 & E2**
One 1½" x 8" (3.8 x 20.3 cm)–**F**
One 2½" x 8" (6.4 x 20.3 cm)–**G**

**OTHER:**

Tube of toothpaste (not brand new and not too empty)–**H**
Thin but strong rubber band–**I**
5 craft sticks–**J**
Toothbrush–**K**
Two 6" (15.2 cm) skewers–**L**
Piece of thick paper, 2" x 2½" (5 x 6.4 cm)–**M**

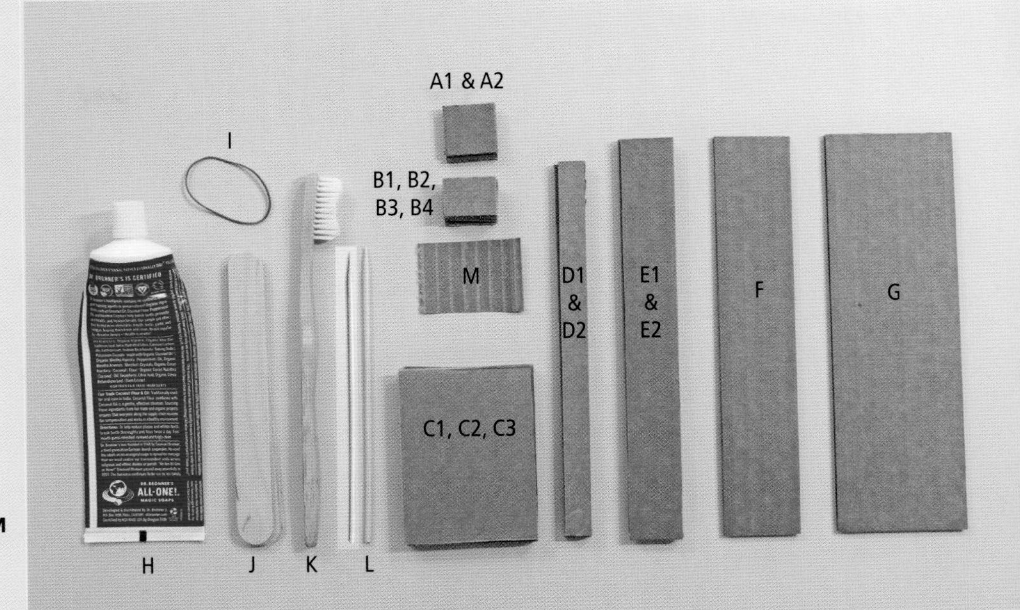


### MAIN STRUCTURE

**CARDBOARD:**

One 1" x 9" (2.5 x 20.3 cm)–**N**
One 2" x 3" (5 x 7.6 cm)–**O**
Two 4" x 4" (10.2 x 10.2 cm)–**P1 & P2**
Two 4" x 7" (10.2 x 17.8 cm)–**Q1 & Q2**
One 4" x 8" (10.2 x 20.3 cm)–**R**
Two 4" x 11" (10.2 x 27.9 cm)–**S1 & S2**

**OTHER:**

6" (15.2 cm) skewer–**T**

*(continued)*

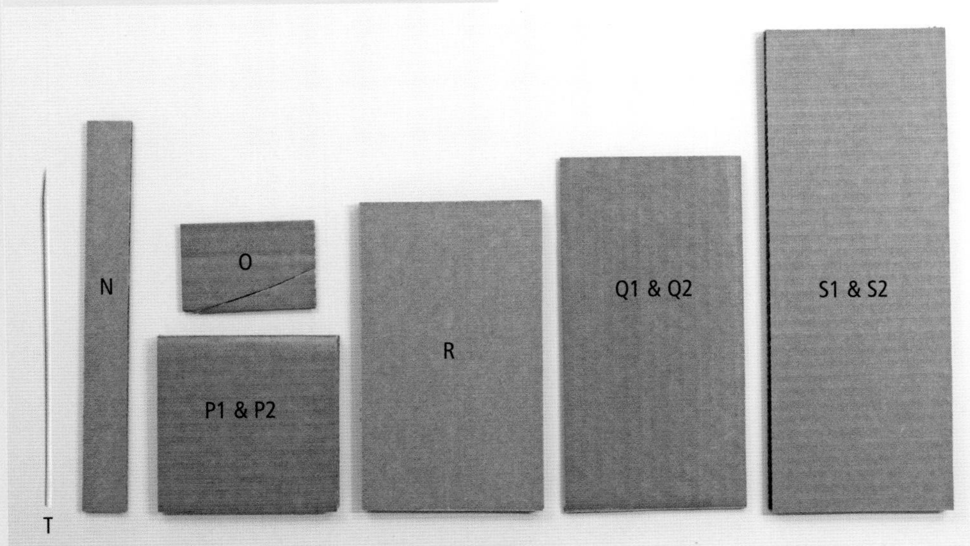


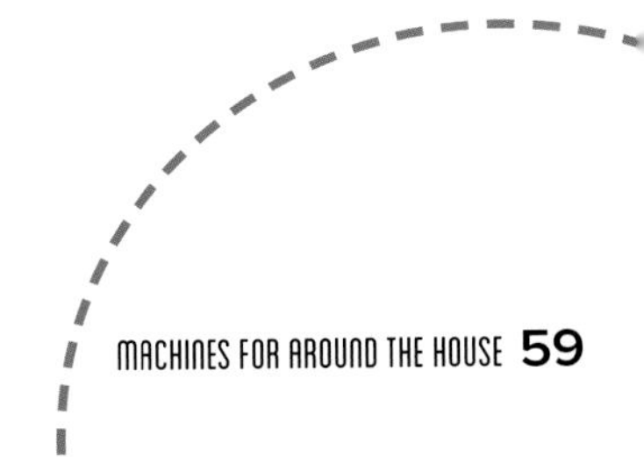

## TOOTHPASTE SQUEEZER

### CARDBOARD:

Two 5" x 5" (12.7 x 12.7 cm)–**U1 & U2**

Two 2" x 12" (5 x 30.5 cm)–**V1 & V2**

One 2" x 5" (5 x 12.7 cm)–**W**

### OTHER:

One 6" (15.2 cm) skewer–**X**

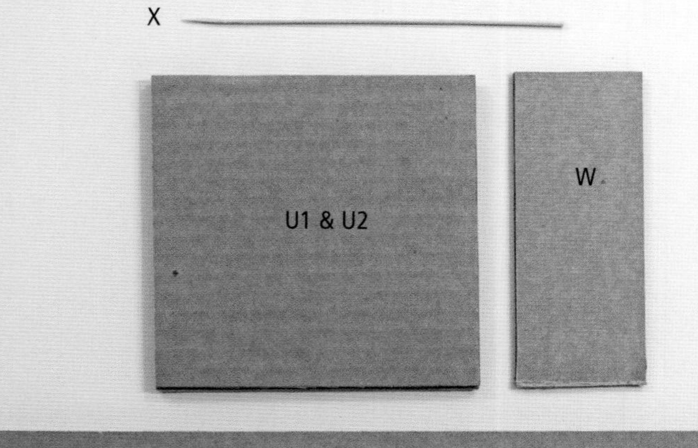


## TAPE RAMP

### CARDBOARD:

Two 2" x 4" (5 x 10.2 cm)–**Y1 & Y2**

Two 3" x 12" (7.6 x 30.5 cm)–**Z1 & Z2**

One 1¼» x 14» (3.1 x 35.6 cm)–**AA**

### OTHER:

One 6" (15.2 cm) skewer–**BB**

Roll of tape–**CC**

3" x 6" (7.6 x 15.2 cm) block or book (½" [1.3 cm] or so thick)–**DD**

Dominoes–**EE**

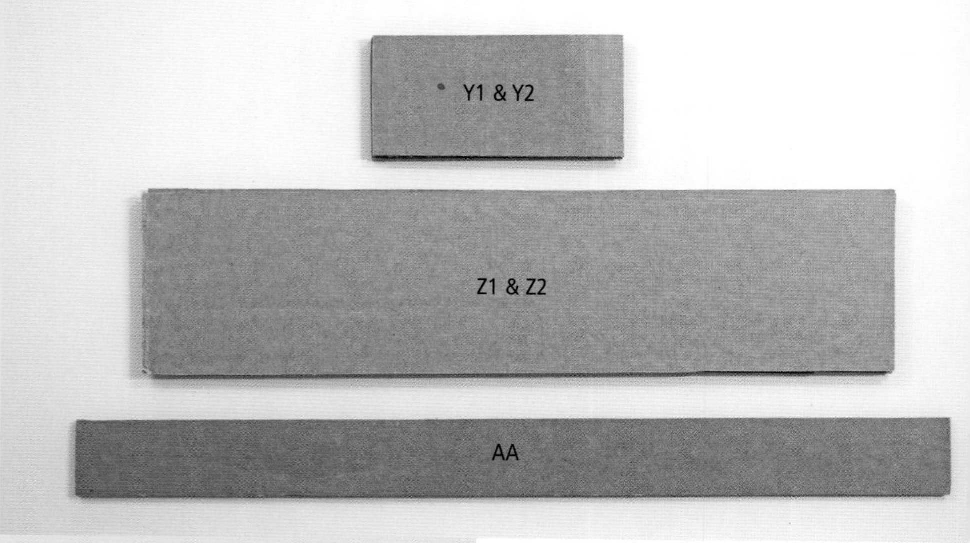


## TOOLS

Scissors

Hot glue gun and glue sticks

Painter's tape

Pencil

Ruler

Awl

Utility knife

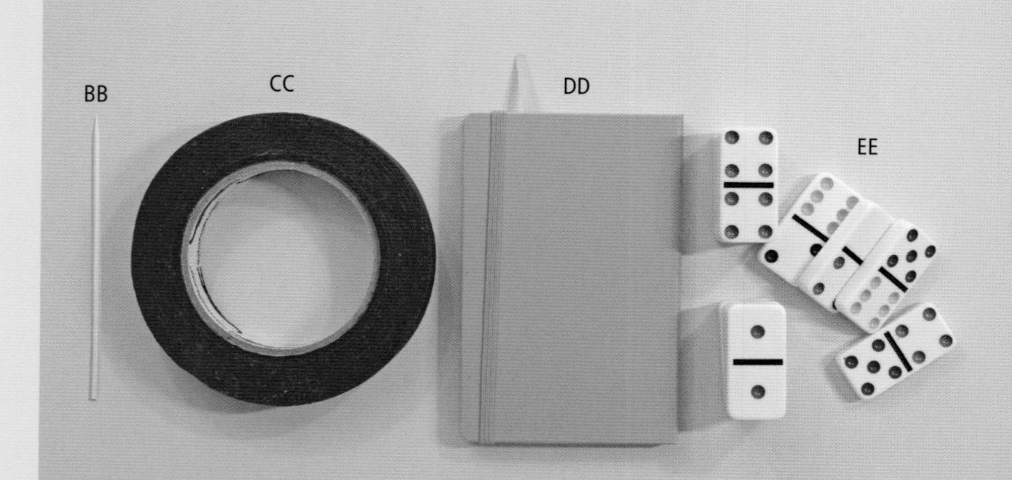

# BUILD THE MACHINE

### Make the Toothbrush Slider

1. Glue D1 and D2 to G **(fig. a)**.
2. Make sure F can move easily back and forth inside the track you created **(fig. b)**. If it's tight, carefully trim it so it can slide freely.
3. Glue a craft stick to each of the D pieces, making sure you don't get any glue on the F piece **(fig. c)**. Also make sure that the sticks are slightly angled upward so F can move freely. The goal is to keep it constrained from left to right but slide with a gentle push.
4. Using tape, create a hinge between E1 and E2 **(fig. d)**
5. Glue that to the top of F **(fig. e)**. It should fit between the craft sticks with a slight gap.
6. Glue B1, B2, B3, and B4 into two stacks of two **(fig. f)**. Glue A1 on top of one stack and A2 on top of the other **(fig. g)**.

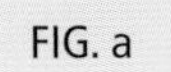

FIG. a

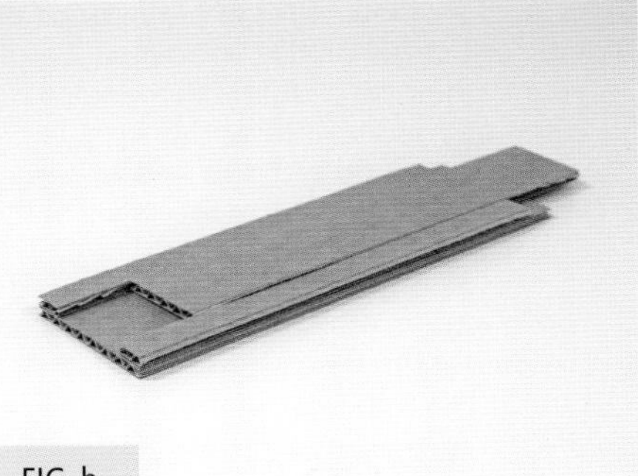

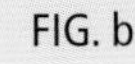

FIG. b

FIG. c

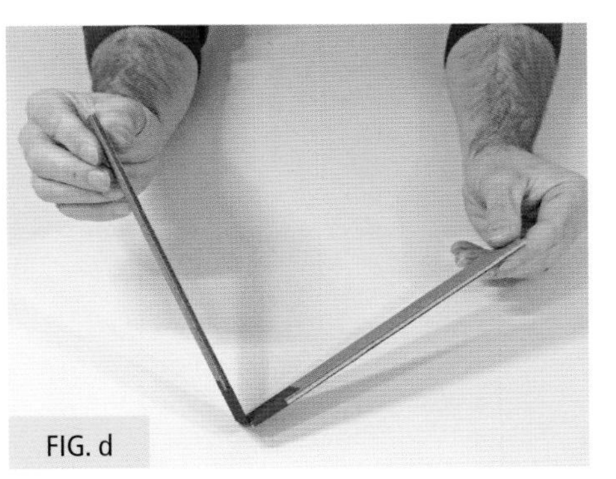

FIG. d

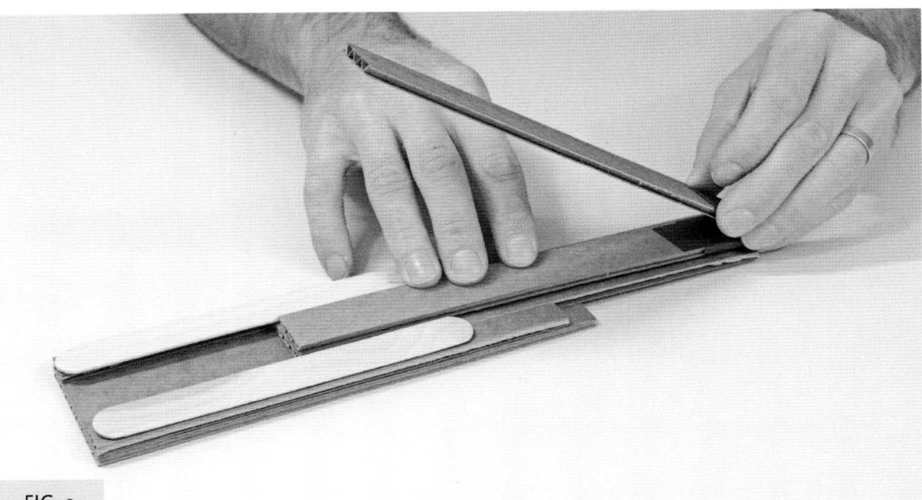

FIG. e

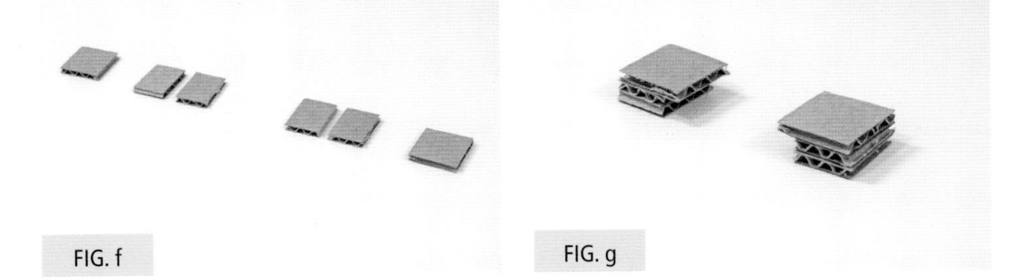

FIG. f

FIG. g

7. Glue the spacers on top of the ends of the craft sticks on the opposite side of your hinged piece. Slide the hinged piece forward slightly and draw a line where it lines up with the A1 and A2 spacers **(fig. h)**. Cut the top layer of the E piece at this line.

8. Using a ruler, draw a line 2" (5 cm) from the end of the hinge and poke a hole through the side of the top hinge piece with an awl. Insert a skewer and cut it so there's ½" (1.3 cm) sticking out on each side **(fig. i)**.

9. Glue C1, C2, and C3 to the sides and front **(figs. j and k)**. Use the top edge of the A1 and A2 spacers as a guide.

10. Cut the rounded portion off one end of a craft stick and glue it to the end **(fig. l)**.

11. Wrap the thick paper (M) around the end of the toothbrush and glue it so it forms a tube. Then glue this to the hinged strip **(fig. m)**. Make sure you don't accidentally glue your toothbrush to anything. It should slide out of the wrapped paper.

12. Tape the rubber band to the craft stick about 1" (2.5 cm) above the top of the cardboard support. Wrap it around the skewer on both sides. Slide your toothbrush back in and test the mechanism to make sure it works. It should easily lift your toothbrush to a nearly vertical position **(figs. n and o)**.

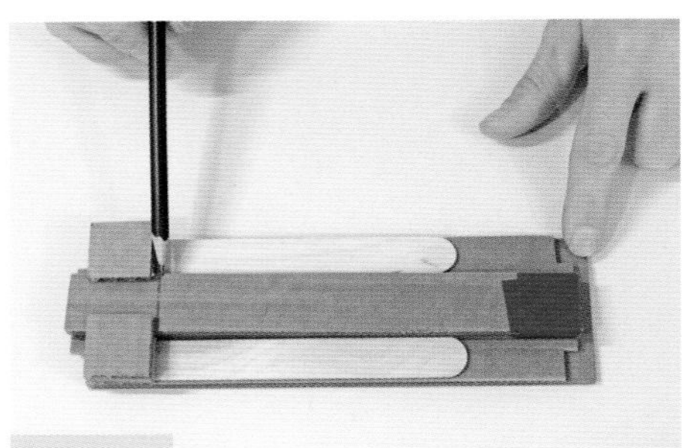
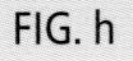
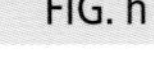

FIG. h

FIG. i

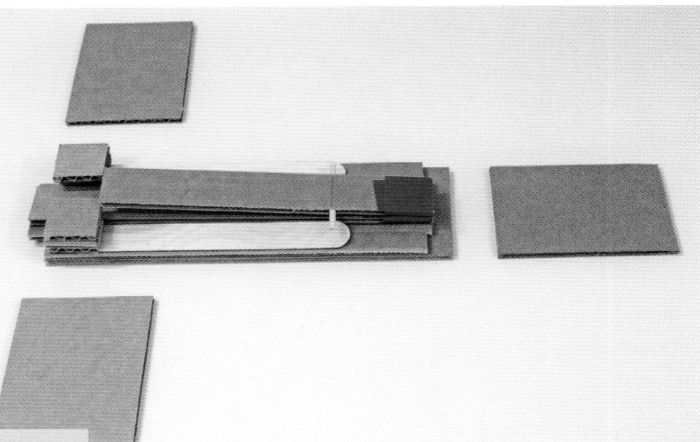

FIG. j

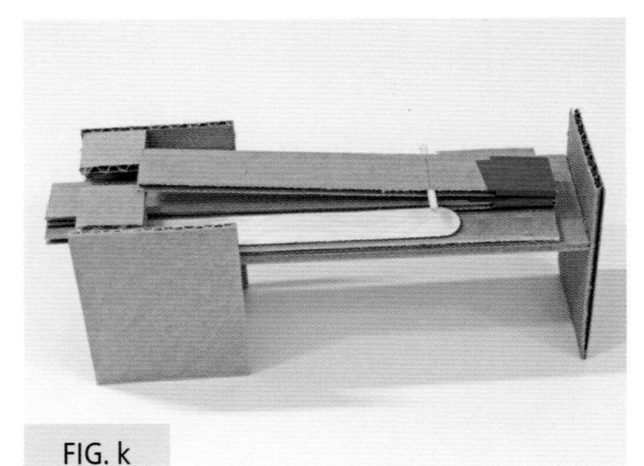

FIG. k

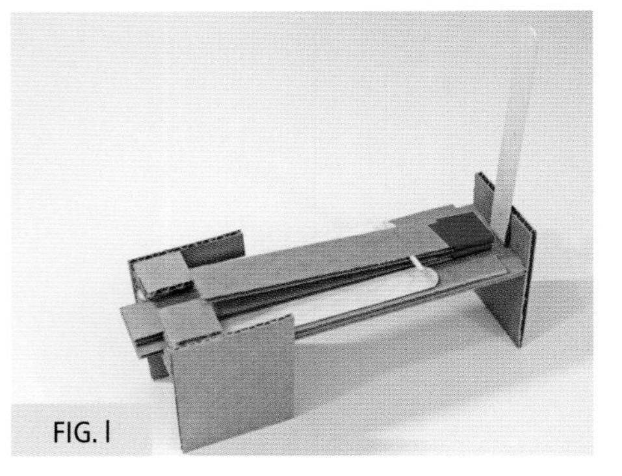

FIG. l

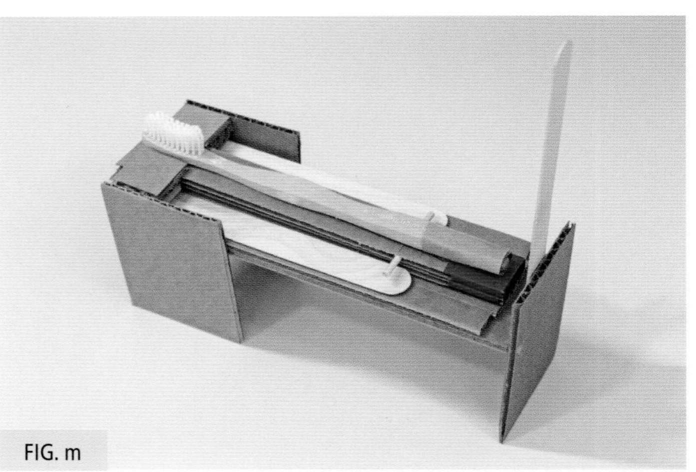

FIG. m

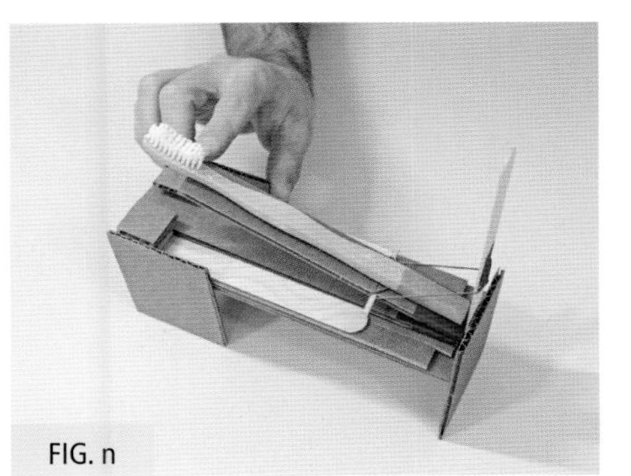

FIG. n

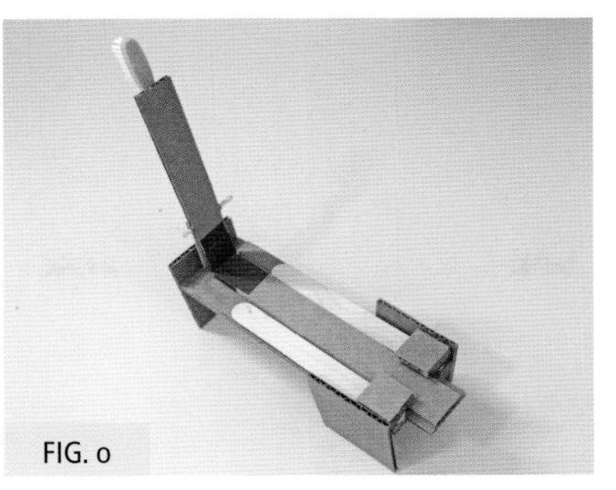

FIG. o

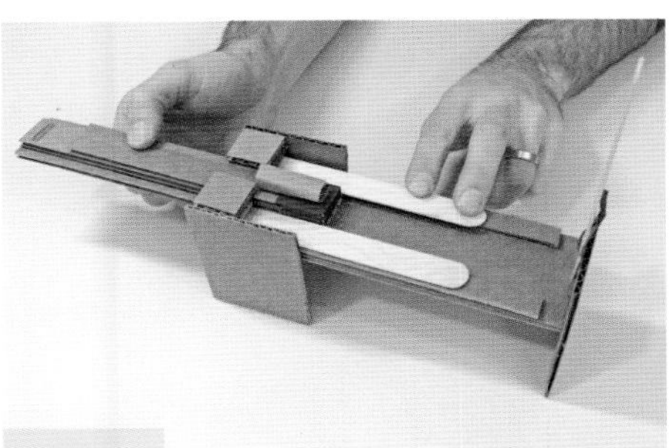

FIG. p

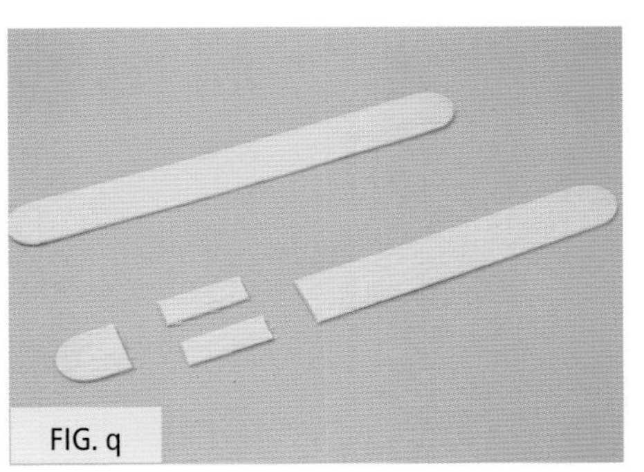

FIG. q

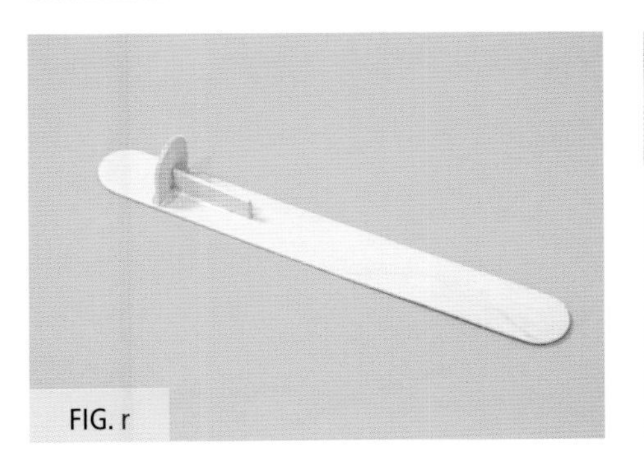

FIG. r

FIG. s

FIG. t

13. Next we'll cut a slit in the back that will act as a stopper. Remove the toothbrush and unhook the rubber band. Pull out the skewer and carefully slide the whole assembly out the back. With a utility knife, cut a slit through both layers, about ¼" (6 mm) from the near. Later on, depending on a lot of factors (toothbrush length, how full your toothpaste tube is, etc.), we might have to recut this slit. Once the slit is cut, put everything back together **(fig. p)**.

14. Take two craft sticks and cut one as shown in **fig. q**. The end piece should be about ¾" (1.8 cm) and the other piece about 1" (2.5 cm) long and then cut in half. Glue the short piece about 1" (2.5 cm) from the end of the other stick and glue the short piece on the other side for support **(fig. r)**.

15. Slide the toothbrush mechanism (with the toothbrush removed) back until the top hinge piece lines up with the A pieces. Hold it here while you position the craft stopper lever so that the ¾" (1.8 cm) curved piece is inside the cut slit **(fig. s)**. Try to keep it held tight while you mark the bottom so you know where to glue the craft stick. Once marked, you can let go of the slider and glue the craft stick in place. Only apply glue to the end of the stick. It needs to have some spring to it, like a diving board.

16. Test the mechanism to make sure it works. Do this by sliding the toothbrush assembly back until the craft stopper lever clicks into the slit. When you push down on the end of the craft stick, it should release the assembly, sending the toothbrush first forward and then upright **(fig. t)**.

## Build the Main Structure

1. Glue together P1 and P2, and cut a slit in the middle that's as long as the end of your toothpaste tube is wide (2½" [6.4 cm] or so) **(fig. u)**.
2. Glue S1 and S2 to each side of the base (R). Attach the doubled-up P pieces to the top, making sure the slit is facing the long end of the base **(fig. v)**.
3. With an awl, poke a hole in both Q1 and Q2 1" (2.5 cm) from the edge and 1½" (3.8 cm) or so from the bottom. This location might have to be altered later, again, depending on how much toothpaste is in your tube. For now, this will be our starting point. Cut a slit from the hole to the edge of the cardboard. This will let us slide a skewer in and out. Glue these pieces to the top of the main body as shown in **fig. w**.

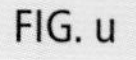

FIG. u

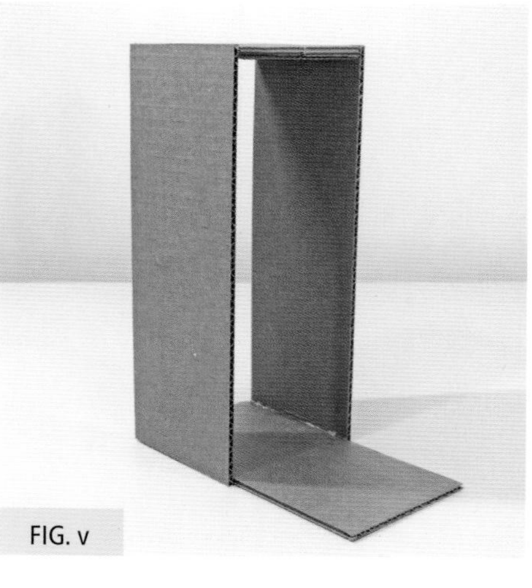

FIG. v

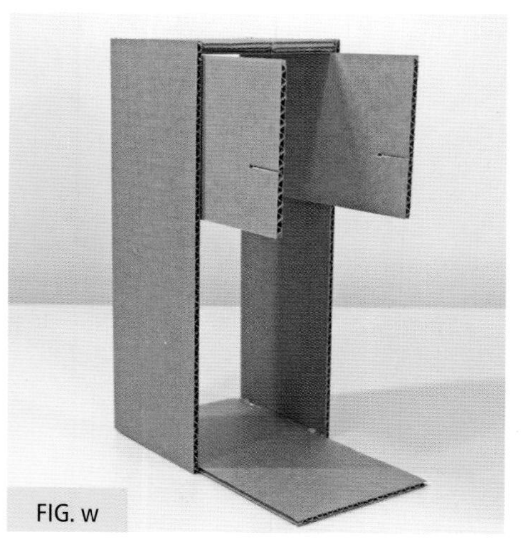

FIG. w

## Make the Toothpaste Squeezer

1. Glue together V1 and V2. Cut a slit 2" (5 cm) from the end that's about ½" (1.3 cm) wide and slightly longer than the width of your toothpaste tube **(fig. x)**. This is what will squeeze your toothpaste tube.
2. Glue and tape a 6" (15.2 cm) skewer to the end (on the slit side) **(fig. y)**.
3. Cut a second skewer so it's as wide as your toothpaste tube. Tape it to the end of your tube **(fig. z)**. This will allow it to hang without sliding out of the slit you cut earlier.

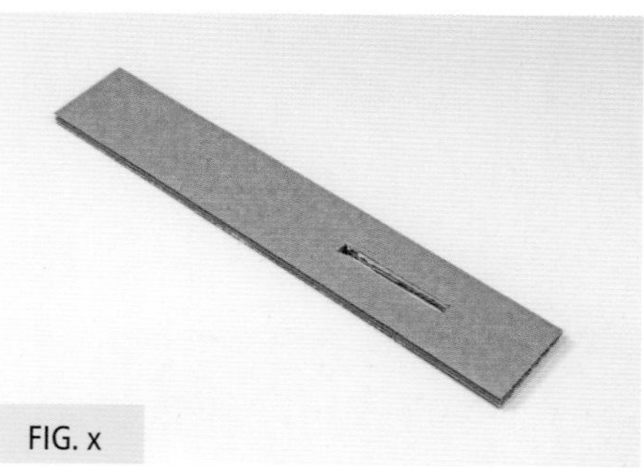

FIG. x

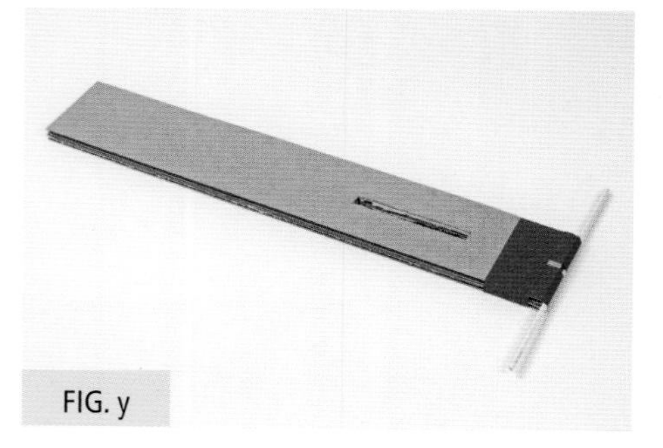

FIG. y

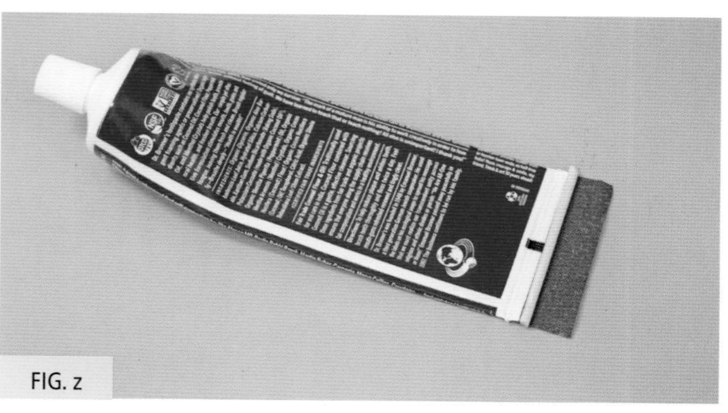

FIG. z

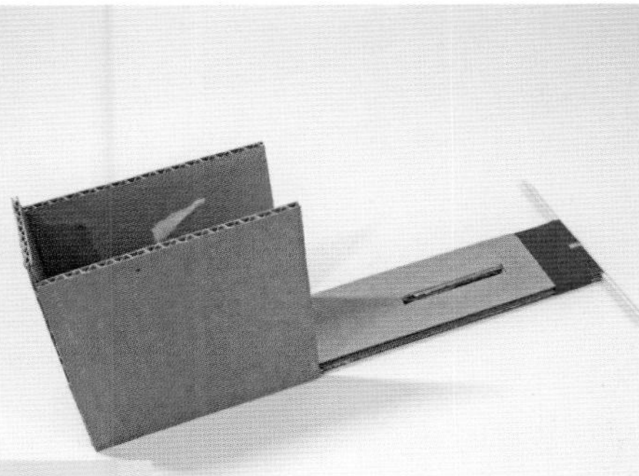
FIG. aa

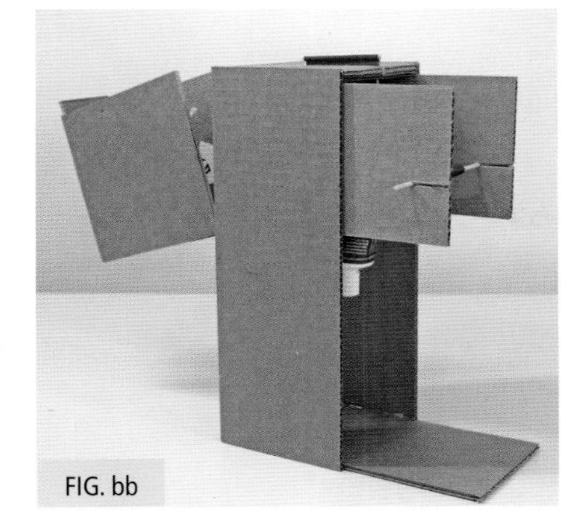
FIG. bb

4. Glue U1 and U2 to the back of the toothpaste squeeze lever. Add W **(fig. aa)**. This is where the roll will fall into. The tape will act as a weight to pull down on the lever, which will squeeze out the toothpaste.

5. This next step might be tricky at first, but after you do it once or twice it's a piece of cake. Slide your toothpaste tube into the large cutout on the toothpaste squeezer lever **(fig. bb)**. Then, in one motion, slide the end of the tube into the slit while simultaneously pushing the skewers into the slits until they pop into the poked holes **(figs. cc and dd)**.

FIG. cc

FIG. dd

### Make the Tape Ramp

1. Cut a 2" x 2" (5 x 5 cm) triangle off one corner of Z1 and Z2 **(fig. ee)**.
2. Glue AA on the short side so you create a channel. Make sure the extra bit is on the cut corner side **(fig. ff)**.
3. Glue the ramp to the top of the main body, with the opening facing the tape holder. Add Y1 and Y2 to the sides for extra support **(fig. gg)**.
4. Now we need to make sure the toothbrush is in the right position. Pull the slide mechanism back until the craft stick goes into the slit. Line up the bristles of the toothbrush with the toothpaste tube **(fig. hh)**.
5. Next, release the craft stick so the toothbrush mechanism can lift. You'll notice that my toothbrush bumps into the skewers. Yours might be shorter than mine and clear it no problem. If not, then you'll have to fix it **(fig. ii)**.
6. If your toothpaste tube is mostly full of toothpaste, you can probably raise the positions of the skewers because you don't need the lever to go down as far to squeeze out the paste. Mine was less than half full, so I need to keep this pivot location where it is **(fig. jj)**.

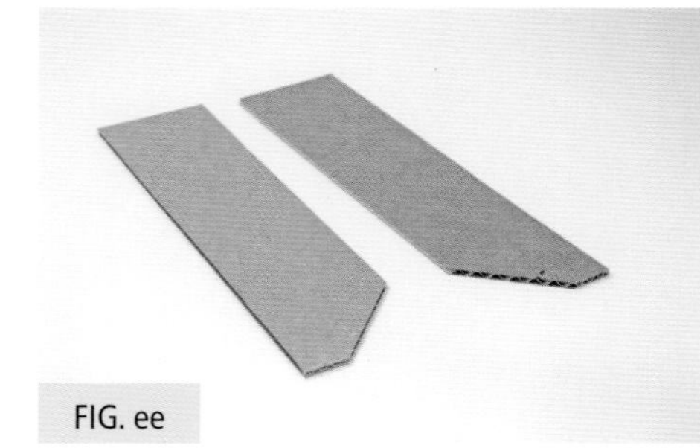
FIG. ee

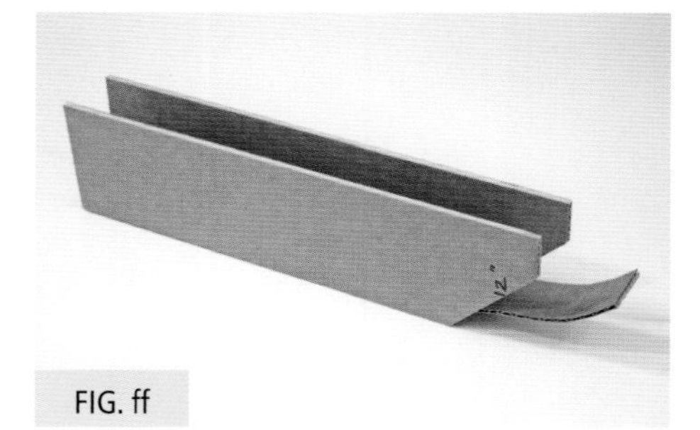
FIG. ff

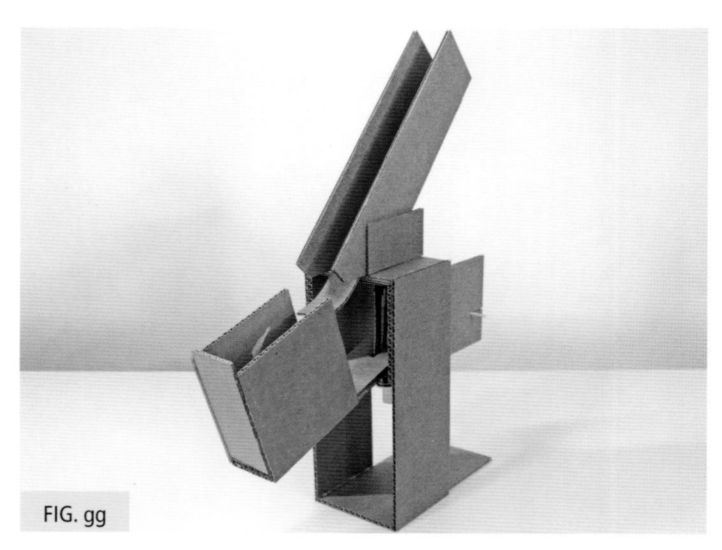
FIG. gg

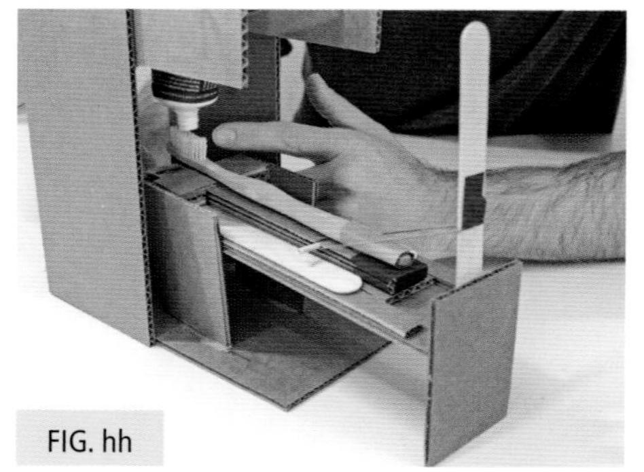
FIG. hh

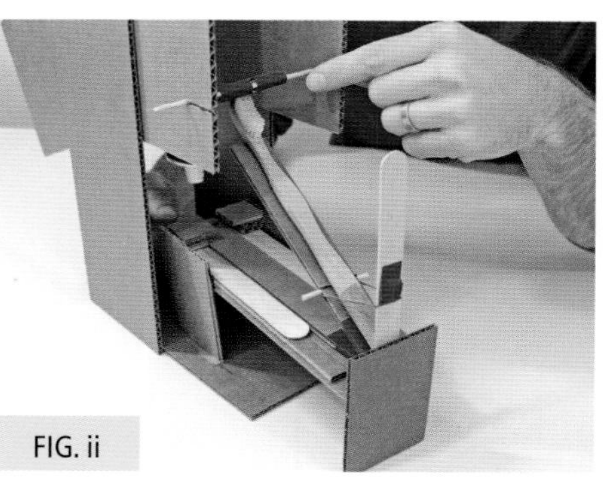
FIG. ii

FIG. jj

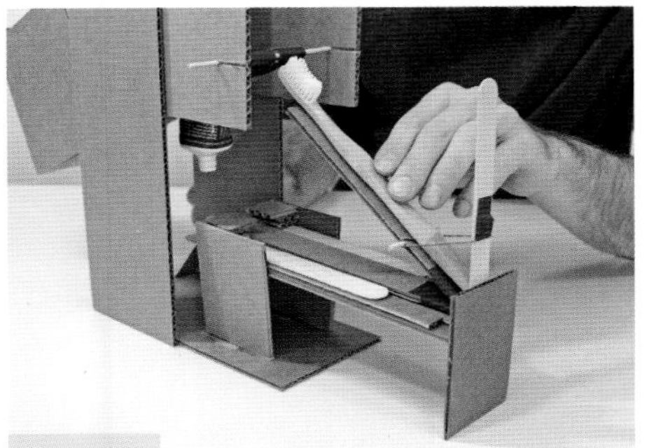
FIG. kk

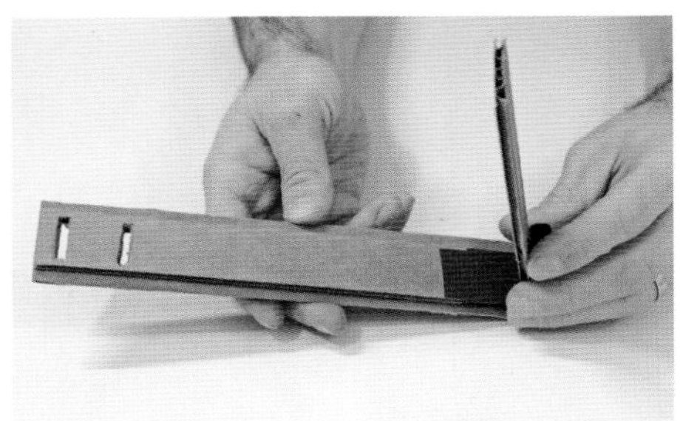
FIG. ll

FIG. mm

FIG. nn

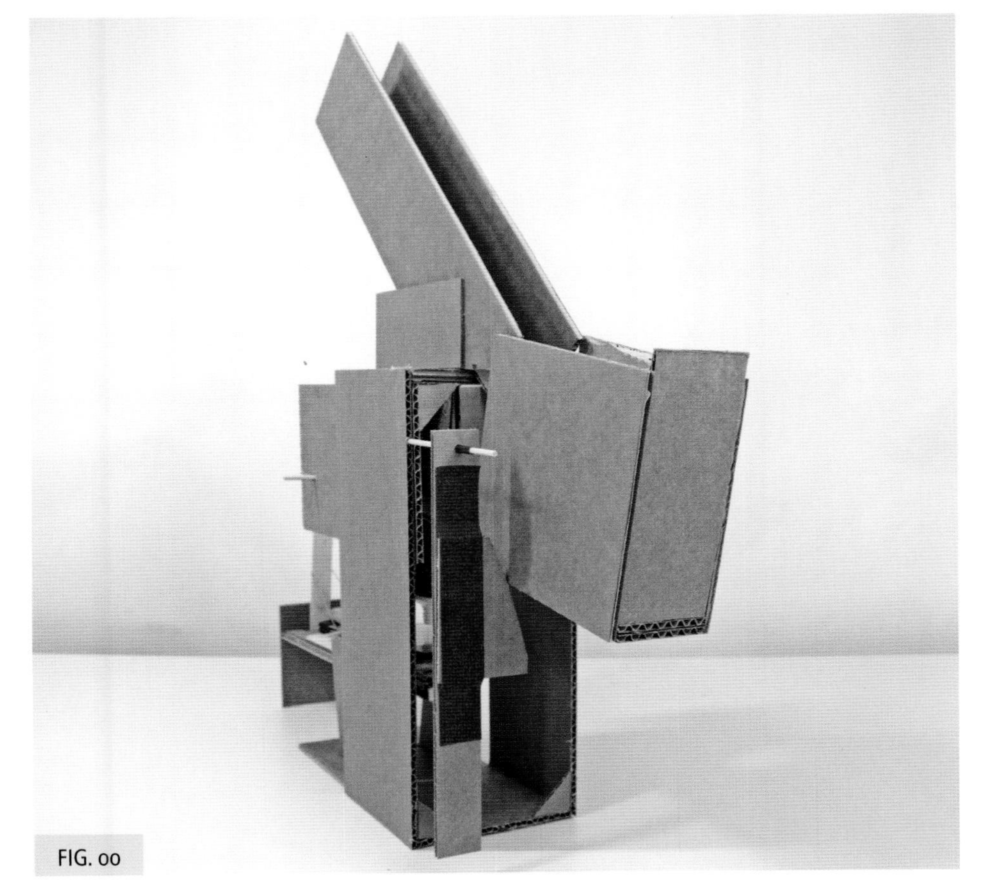
FIG. oo

7. Figure out where the toothbrush needs to be in order to clear the skewers. Once that's established, you can alter the slit location to make everything work. Just mark where the slit needs to be in order for the toothbrush bristles to line up with your tube. Mark this spot and then cut a new slit **(figs. kk and ll)**.

8. Because I had to change the location of my toothbrush mechanism, I needed to add an extender to the craft stick lever. I wanted it to be about 1" (2.5 cm) or so in from the back edge of the main structure **(fig. mm)**.

### Attach the Domino Lever

1. Cut a curve into O and poke a hole ½" (1.3 cm) from the top edge of N. Insert a skewer into the hole and poke this into the main body so the bottom of the skinny piece is about ½" (1.3 cm) from the table top. Use tape to attach the curved piece to N just below the box that's attached to the lever **(figs. nn and oo)**.

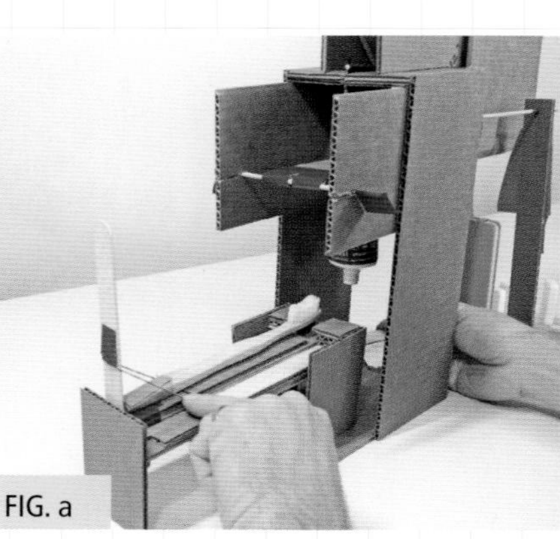
FIG. a

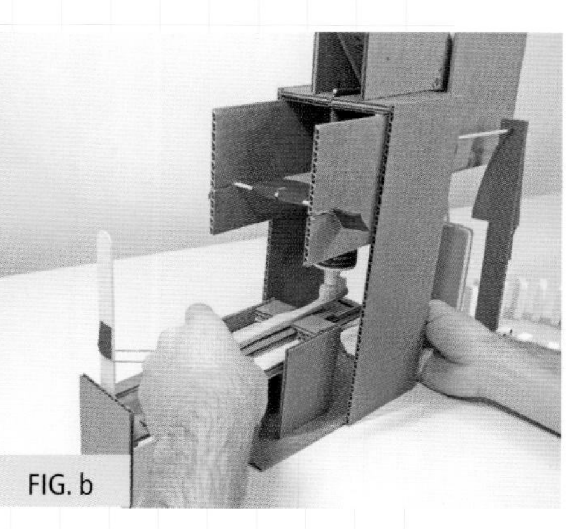
FIG. b

# START THE CHAIN REACTION!

FIG. c

FIG. d

### The Setup

1. To set your toothbrush, push it toward the rear of the machine (by holding the skewer; if you grab your toothbrush it will slide out of its holder) and push down on the craft stick lever. Once the slit is lined up, release the lever so it engages the slit cutout **(figs. a and b)**.
2. The small notebook or piece of wood will act as a large domino. Place this in a way so that, when it falls, it will trigger the craft stick lever, which will release your toothbrush. Place dominoes in a continuous loop from the edge of the domino lever to the notebook. You'll need to build up to the notebook because one domino won't be heavy enough to knock it over. I went with a 1-2-3 combination and that seemed to work well **(figs. c, d, and e)**.

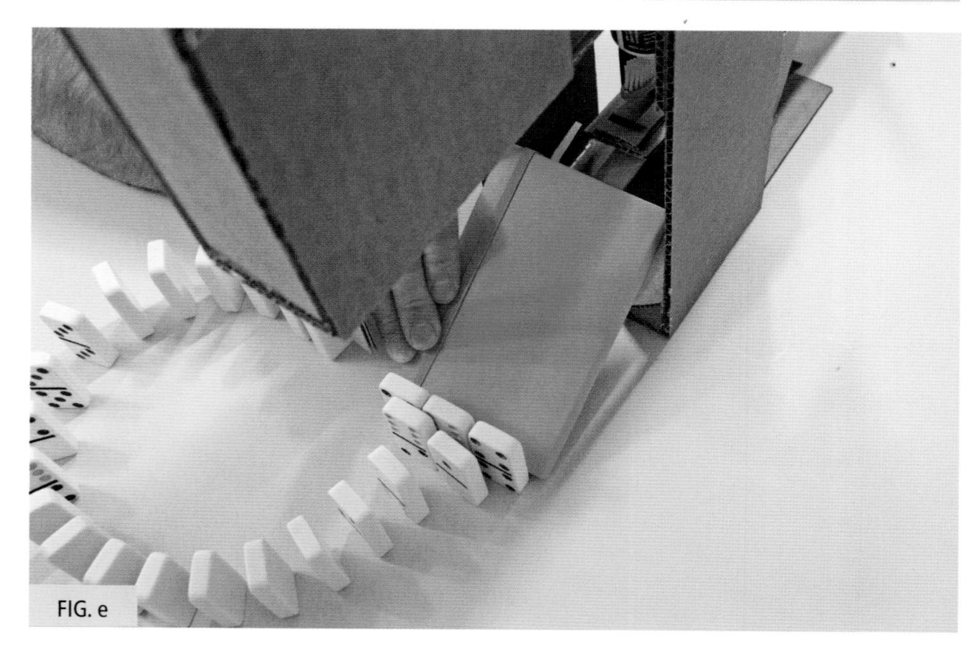
FIG. e

## Make It Go!

To set off the machine, roll the tape down the steep ramp. When it falls into the box, the weight of it will pull down on the lever, squeezing out a small portion of toothpaste. At the same time, it will set off the dominoes by falling onto the domino lever. The dominoes act as a sort of timer. They give the toothpaste time to squeeze out and fall onto the bristles of your toothbrush. The longer the chain of dominoes, the more time gravity has to pull down on the toothpaste.

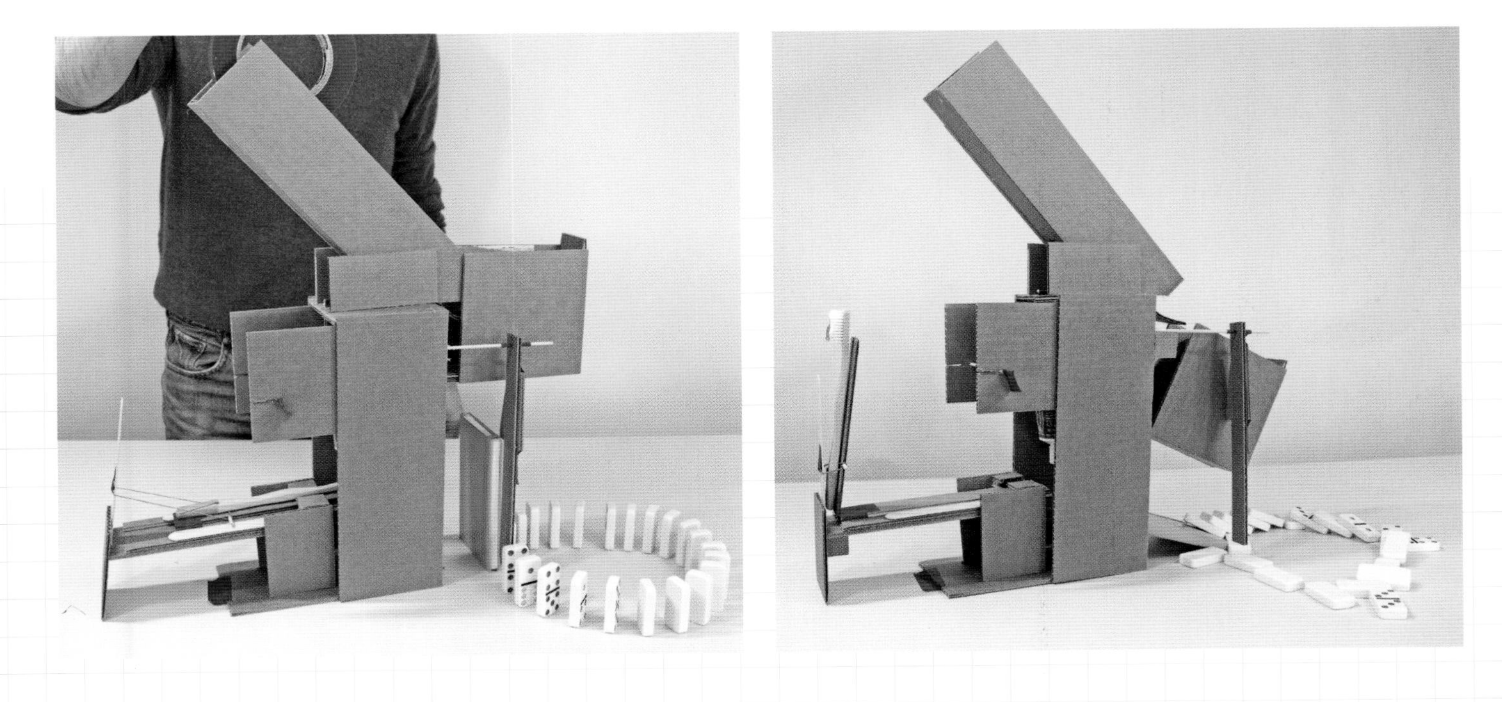

## Troubleshooting

The most common issue you might have is that your sliding mechanism isn't smooth enough. You can press on the edges of the pieces with your fingers to try and flatten them out so they glide more smoothly under the rails.

An issue I ran into was that my toothpaste tube wasn't quite full enough. A tiny amount would come out, but not really enough to make it satisfying. Instead of grabbing a new tube of toothpaste, I added some weight to my roll of tape in the form of pennies. I taped some cardboard on the front and back of the roll to keep them from falling out **(fig. a)**. This gave me more force to push on the lever, which helped squeeze out more of the paste.

The front of my machine became a little wobbly, so I added a cardboard support to stiffen it up **(fig. b)**.

I also added a support to the skewer that holds the domino lever because it was moving around too much **(fig. c)**.

If you don't have a small notebook, a thin piece of wood will do the trick as well. You could also use cardboard with weights attached to it, like pennies or marbles **(fig. d)**.

### Engineering Know-How

When the toothbrush slider is engaged, the rubber band has potential energy. As the lever is pushed down by the book, it disengages the slider. The potential energy from the rubber band now becomes kinetic energy. Since the rubber band is mounted above the hinge of the toothbrush slider, it pulls it forward and also raises it. If it was mounted even with the hinge, it would only cause it to slide forward without raising it as well.

FIG. a

FIG. b

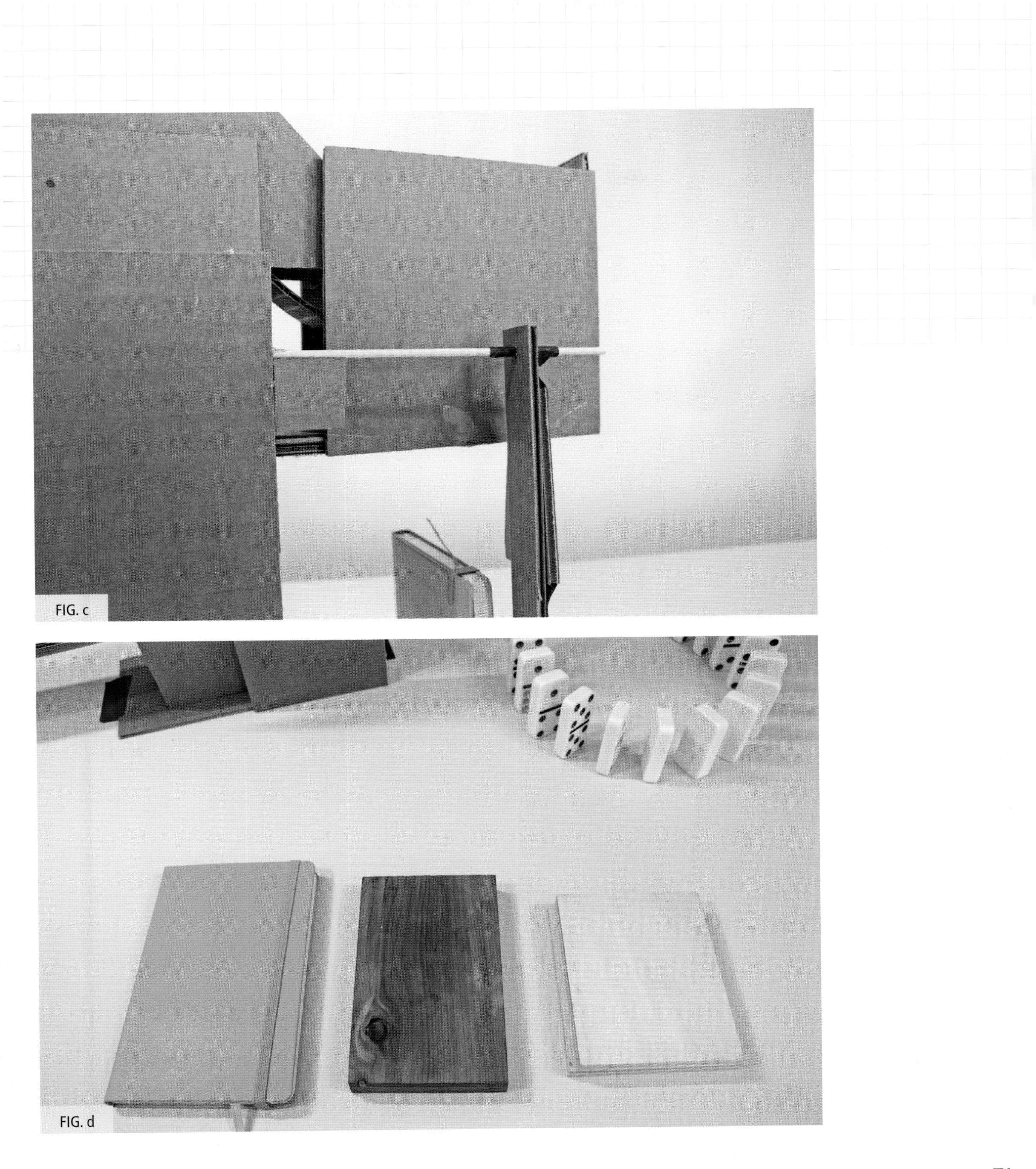

FIG. c

FIG. d

# SOAP DISPENSER

Lots of germs going around these days. The best way to not get sick is to wash those hands. Sure, you could just use a boring bar of soap, or pump it from the dispenser with your hand. Or, you could build this chain reaction soap dispenser to do it for you.

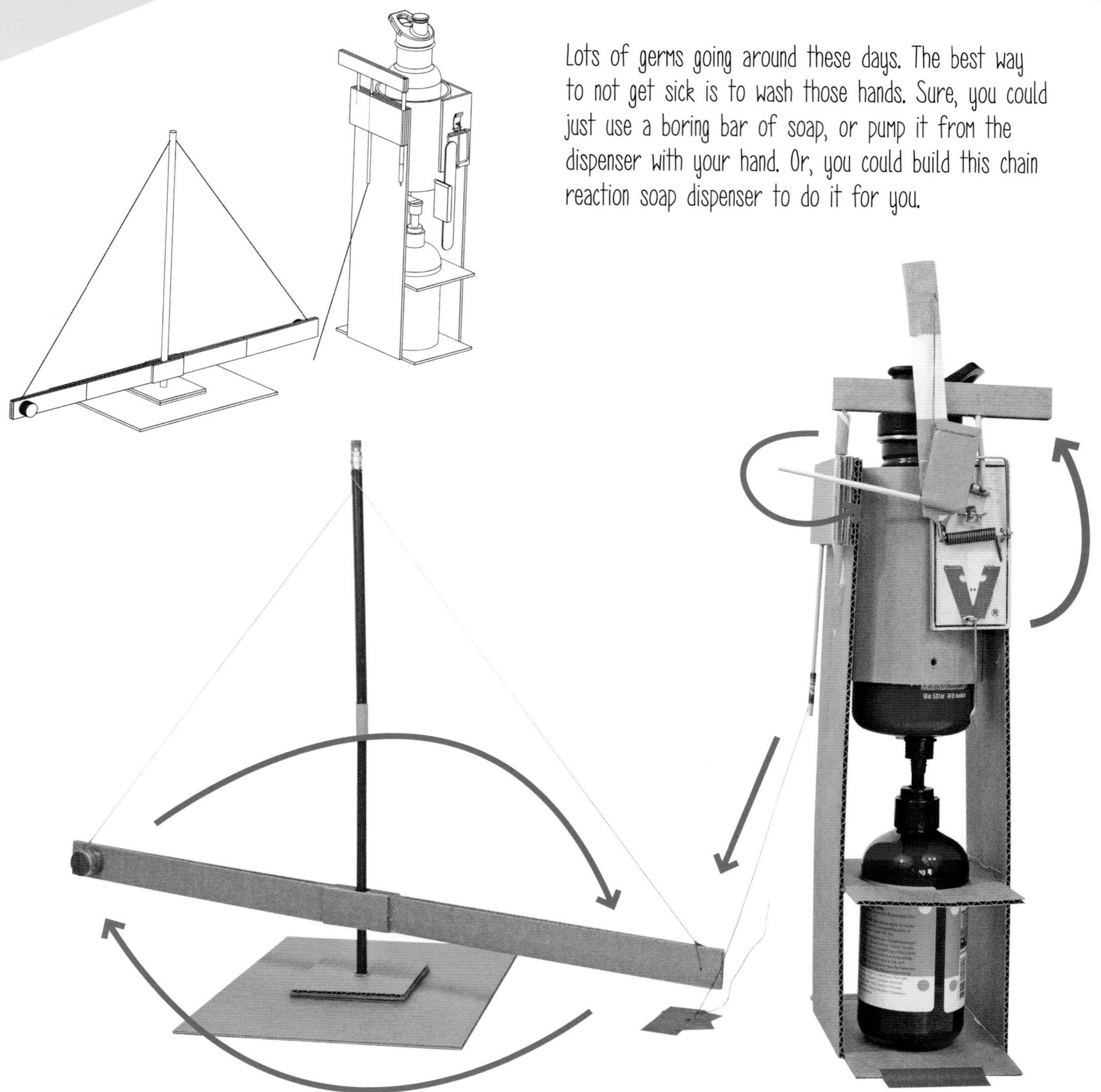

# BUILD MATERIALS & TOOLS

## CANTILEVER SWING ARM

### CARDBOARD:

Two 3" x 3" (7.6 x 7.6 cm)–**A1 & A2**
One 7" x 9" (18 x 23 cm)–**B**
Two 1" x 2" (2.5 x 5 cm)–**C1 & C2**
Two 1" x 1" (2.5 x 2.5 cm)–**D1 & D2**
Four 1" x 10" (2.5 x 25.4 cm)–**E1, E2, E3 & E4**

### OTHER:

40" (100 cm) string–**F**
16 coins or washers for weight–**G**
2 new pencils or 15" (38 cm) dowel–**H**

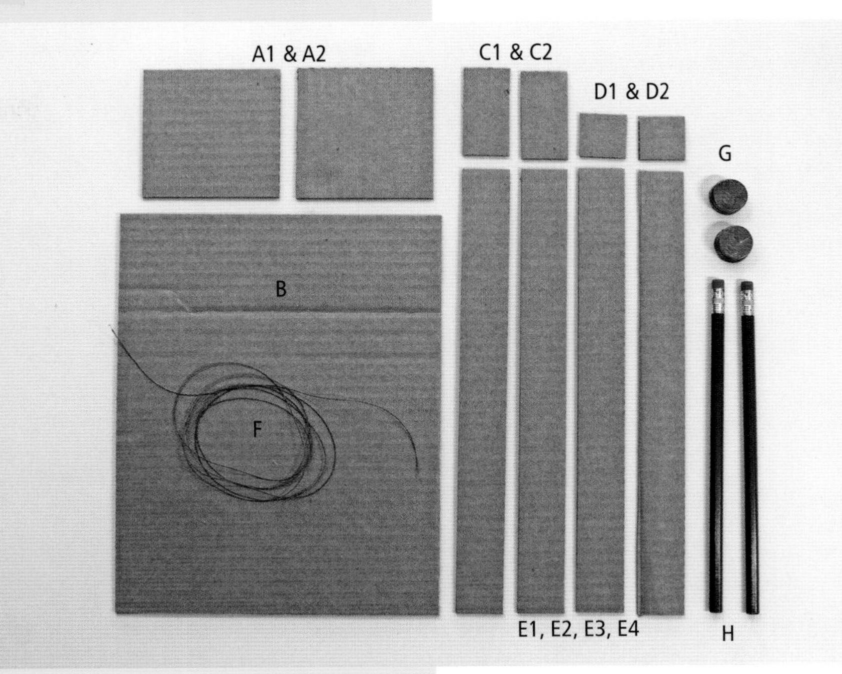


## SOAP DISPENSER BODY

### CARDBOARD:

Two 3½" x 14" (9.5 x 35.6 cm)–**I1 & I2**
One 3¾" x 5" (9.5 x 12.7 cm)–**J**
One 3¾" x 6" (9.5 x 15.2 cm)–**K**
One 1" x 4" (2.5 x 10.2 cm)–**L**
Two ¾" x 5" (2 x 12.7 cm)–**M1 & M2**
One 5" x 10" (12.7 x 25.4 cm)–**N**
Three 2" x 4" (5 x 10.2 cm)–**O1, O2 & O3**

### OTHER:

Mousetrap (Use the one with a metal pin holder, not the fake plastic cheese.)–**P**
Rubber band–**Q**
40" (100 cm) string–**R**
Three 6" (15 cm) skewers–**S**
6" (15 cm) craft stick–**T**
Bottle of soap–**U**
Smooth water bottle (to be used as a weight)– **V**

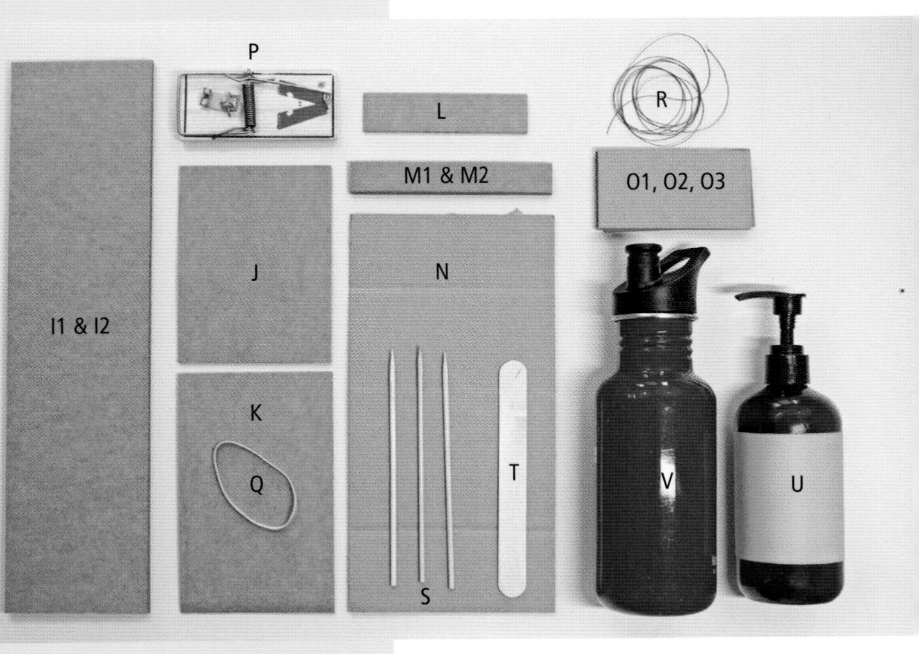


## TOOLS

Hot glue gun and glue sticks
Awl
Sharpened pencil
Pliers
Tape
Utility knife
Ruler
Marker
Scissors
Painter's tape

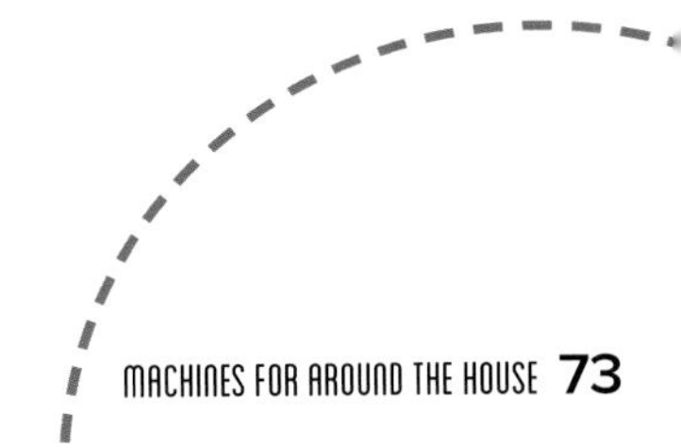

# BUILD THE MACHINE

### Build the Cantilever Swing Arm

1. Glue together A1 and A2, and poke a hole in the center with an awl. Enlarge it with a sharpened pencil **(fig. a)**.
2. Glue this piece to the center of the base B **(fig. b)**.
3. Remove the eraser from one of the pencils. Scrape it out as best you can with a sharp object or the end of pliers. Insert the other pencil into the cavity you carved out. Use glue and tape to secure it **(fig. c)**. (Or use a 15" [38 cm] dowel.)
4. Glue the pencil structure to the base. Make sure the pencils stay as straight and vertical as possible while the glue dries **(fig. d)**.
5. Grab E1 and E2, and lay them out with a ⅜" to ½" gap, a little larger than your pencil **(fig. e)**.
6. Glue C1 so it spans across the longer pieces **(fig. f)**.
7. Flip over and add the D1 and D2 spacers on either side **(fig. g)**.
8. Add another layer of E3 and E4 **(fig. h)**.

FIG. a

FIG. b

FIG. c

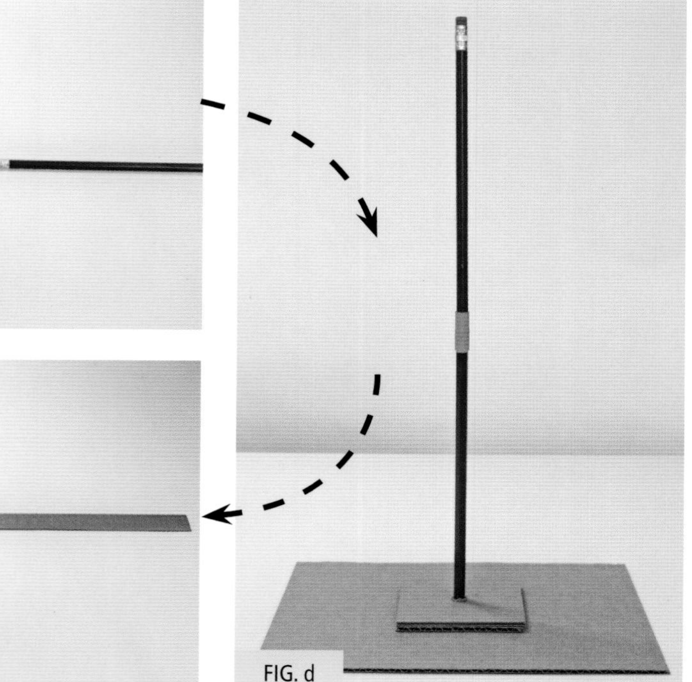
FIG. d

FIG. e

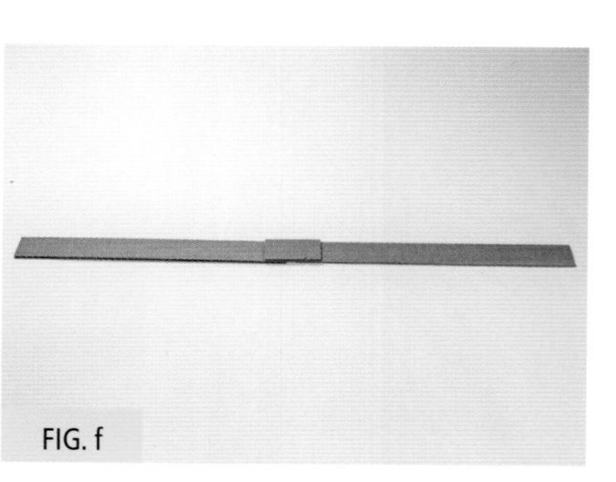
FIG. f

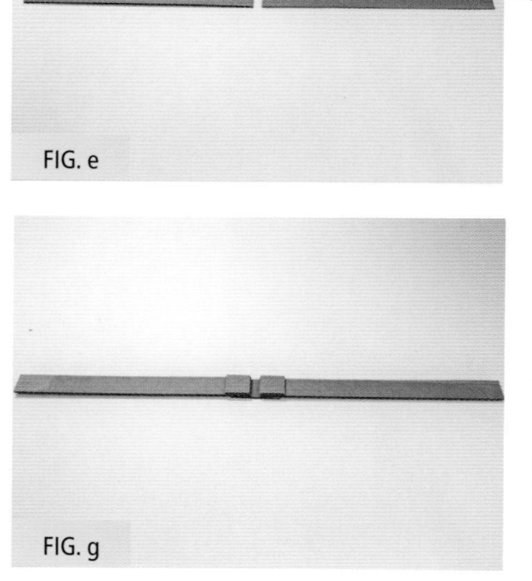
FIG. g

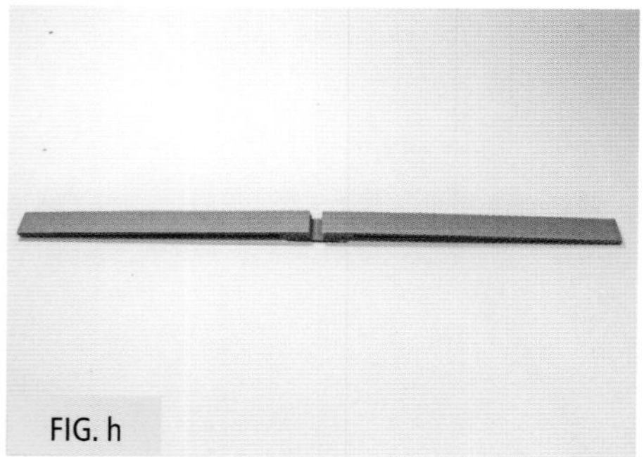
FIG. h

FIG. i

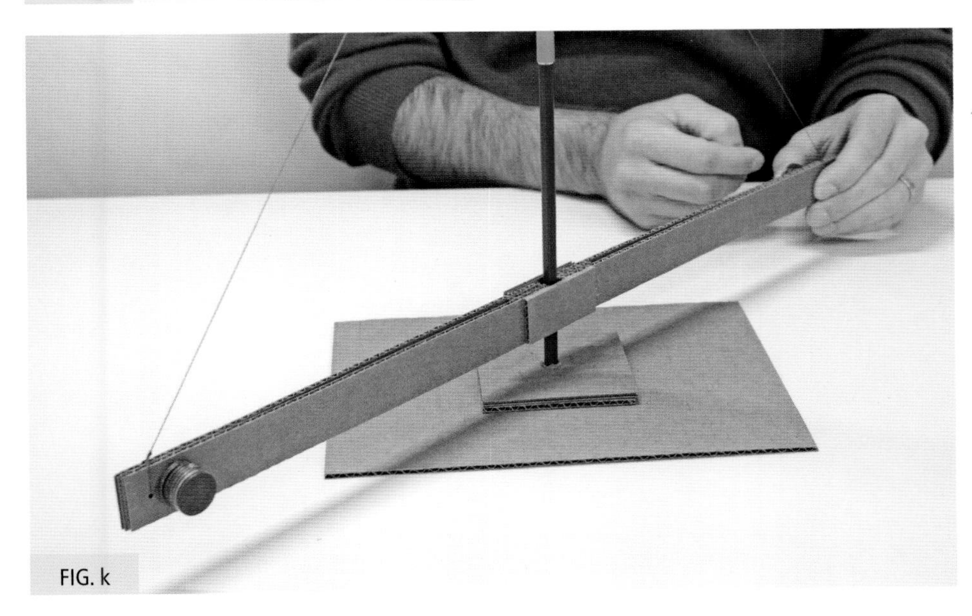
FIG. k

9. Finish it off by adding a C2 spanner. Poke holes about ½" (1.3 cm) from the ends.

10. With a utility knife, cut a slit in the eraser at the top of the pencil structure. Tie the string to one end of the cantilever beam. Slip the string into the slit in the pencil and lift the tied end about 1" (2.5 cm) from the base. Level out the beam and tie off the other end. Slide the string so your beam is as level as possible **(fig. i)**.

11. Make two stack of pennies, about eight in each stack. You can use washers or anything flat with a bit of weight **(fig. j)**. Fasten each stick with glue.

12. Glue the weight to the ends of the beam, making sure they're not facing the same direction **(fig. k)**.

FIG. j

### Make the Soap Dispenser Body

1. Measure the diameter of your soap bottle, and cut a hole that size in the center of J **(fig. l)**.
2. Glue I1 and I2 to either side of K, making sure they're centered on the base **(fig. m)**.
3. Using your soap bottle as a guide, glue the soap holder to the main structure at a height that's just below the rounded part of your bottle **(fig. n)**.
4. Loosely wrap N around your bottle. You want it to be able to slide freely. Glue the piece of cardboard into a cylinder **(fig. o)**.
5. Glue the cylinder to the top of the main body **(fig. p)**.
6. Glue together O1, O2, and O3 **(fig. q)**.

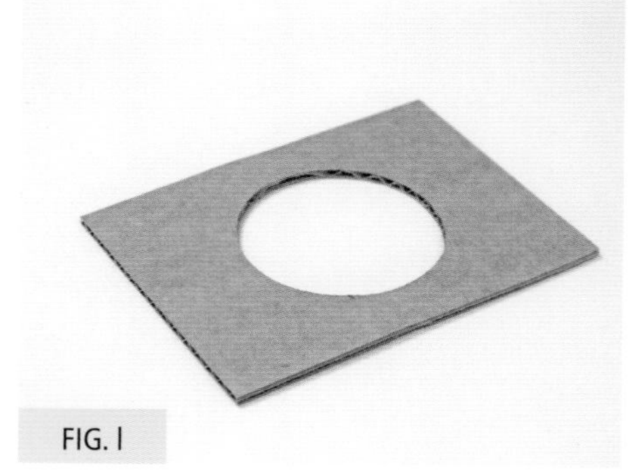

FIG. l

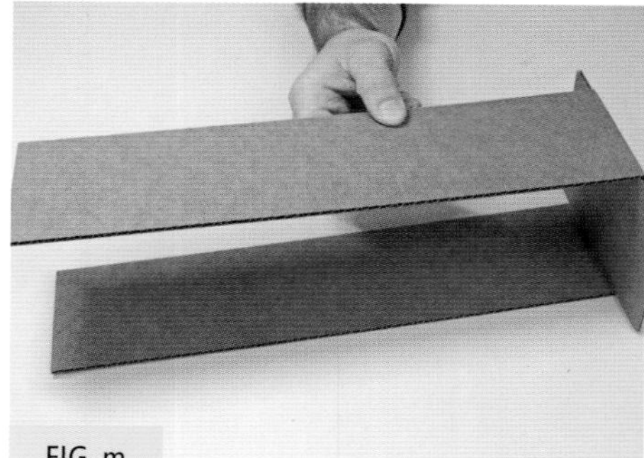

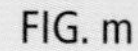

FIG. m

FIG. n

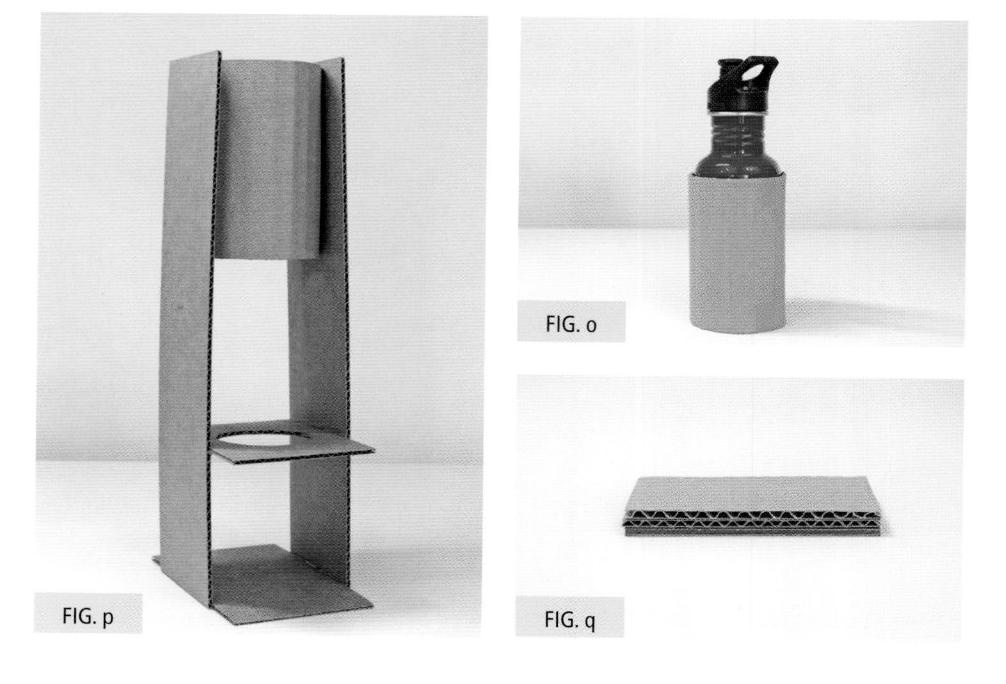

FIG. p

FIG. o

FIG. q

FIG. r

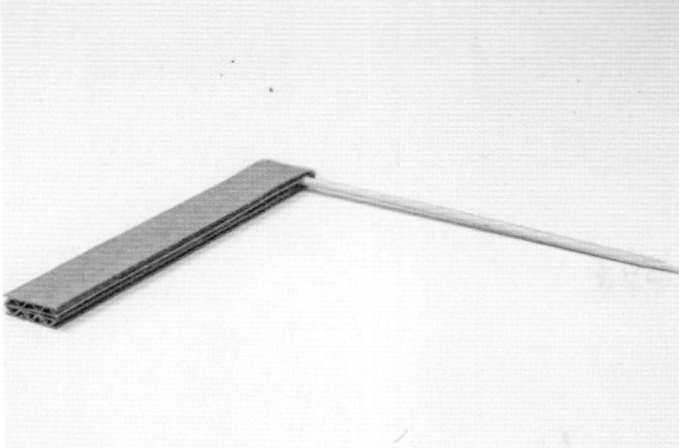
FIG. s

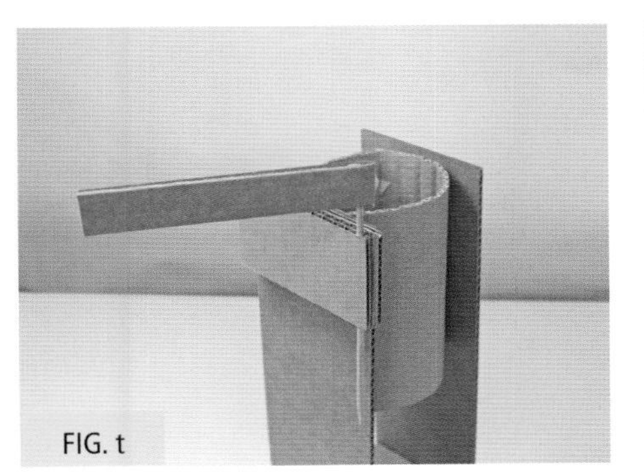
FIG. t

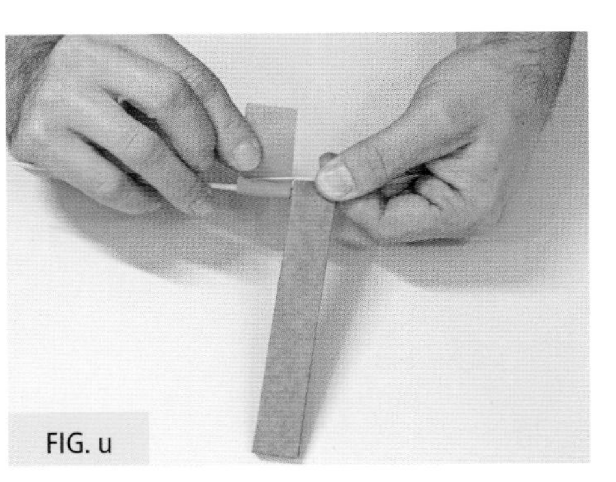
FIG. u

FIG. v

7. Attach them to the front left edge as shown **(fig. r)**.

8. Take one of the skewers and M1 and M2. Glue the skewer between the two pieces of cardboard at the far end **(fig. s)**.

9. Slide the skewer into a corrugated slot of the O stack near the edge. If the skewer doesn't rotate freely you can use a knife to shave off material until it goes back and forth smoothly **(fig. t)**.

10. Wrap some tape around the rubber band and the skewer, just below the cardboard arm. Once you have a bit of it wrapped and it feels secure, pull the rubber band down toward the end of the skewer and wrap more tape around the rubber band so the two strands of rubber are coming out the top edge of the tape **(fig. u)**.

11. Insert the skewer back into its place and add some tape to the skewer just below the cardboard edge. Stretch the rubber band to the bottom and add tape in a similar fashion to the top. The key is to have the ends of the rubber band not be in direct contact with the cardboard, but to be stretched between the far ends of the skewer **(fig. v)**.

### Attach the Mousetrap Mechanism

1. Glue and tape a craft stick to a mousetrap in the location shown **(fig. w)**.
2. Glue the mousetrap to the front of the bottle holder and uprights, using the edge of the right upright as a guide **(fig. x)**.
3. Poke a hole through the bottle holder about ½" (1.3 cm) from the bottom. Insert a skewer and wiggle it slightly so it slides in and out easily. With a bit of skewer sticking out the back, draw a line on the front of the skewer ½" (1.3 cm) from the edge of the bottle holder. Cut on this line. Essentially, you want the skewer long enough to attach a piece of cardboard to one end and still be able to go through both holes on the bottle holder, but not be so long that there's extra skewer sticking out the end **(fig. y)**.
4. Fold L in half. Unfold and place the cut skewer into the fold ½" (1.3 cm) in and glue it in place **(fig. z)**.
5. Then fold the cardboard again and add tape to secure it. Cut small slits near the ends of the cardboard and tie a piece of string to it. Leave about 12" (30.5 cm) or so of string because we need a bit to attach it to the craft stick on the mousetrap **(fig. aa)**.

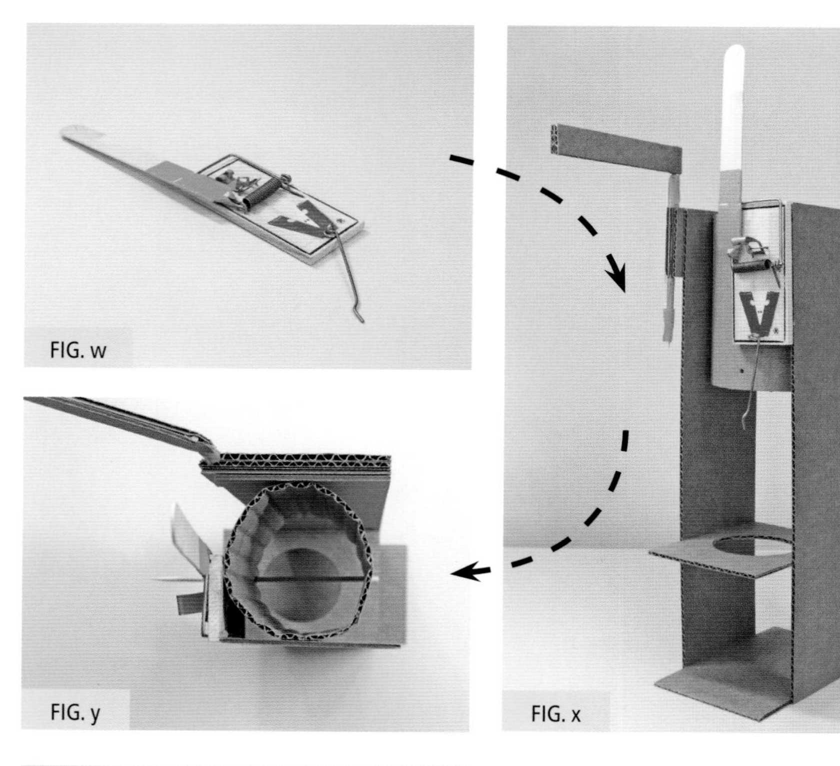

FIG. w

FIG. y

FIG. x

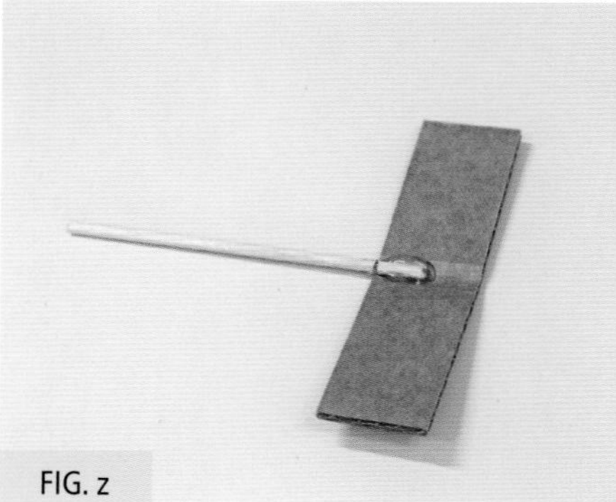

FIG. z

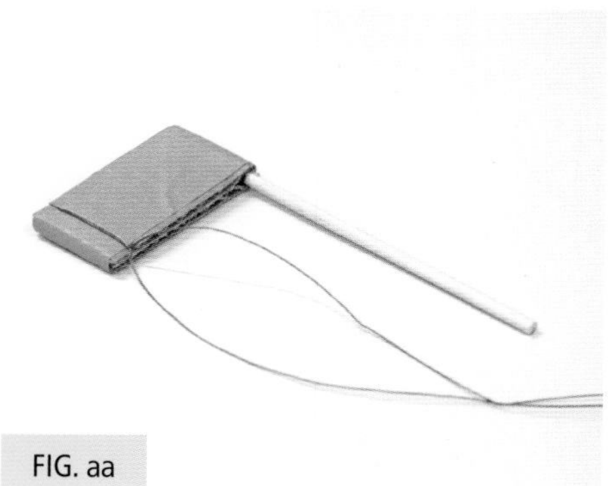

FIG. aa

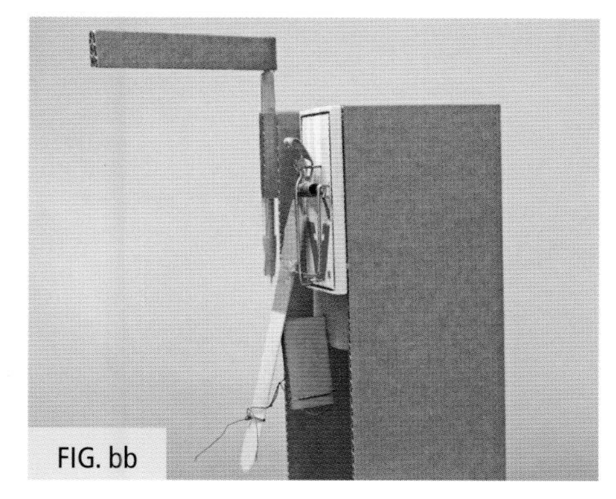
FIG. bb

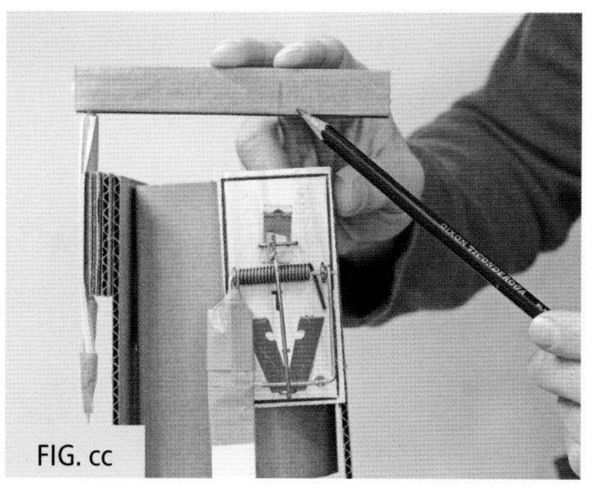
FIG. cc

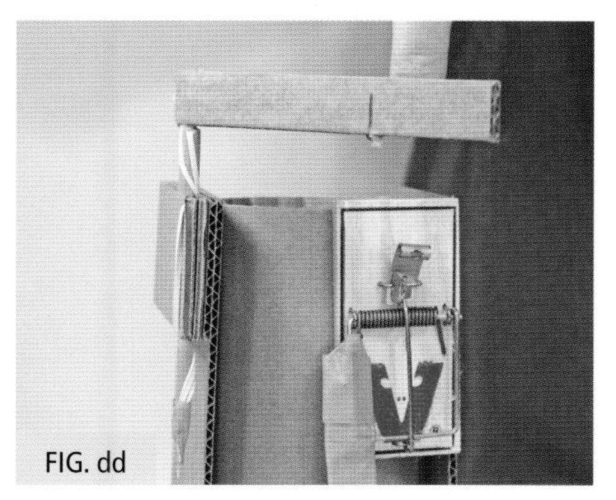
FIG. dd

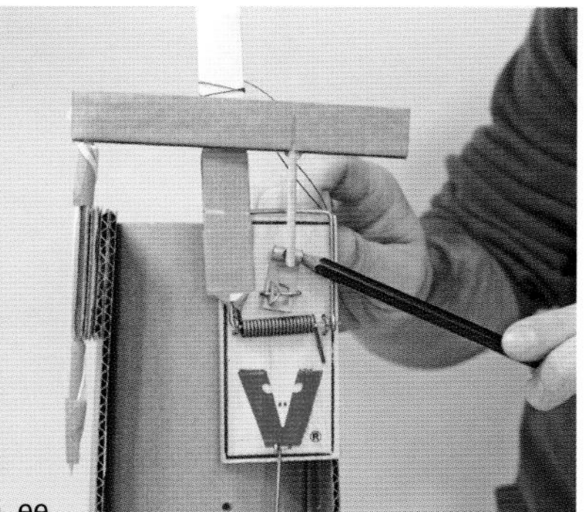
FIG. ee

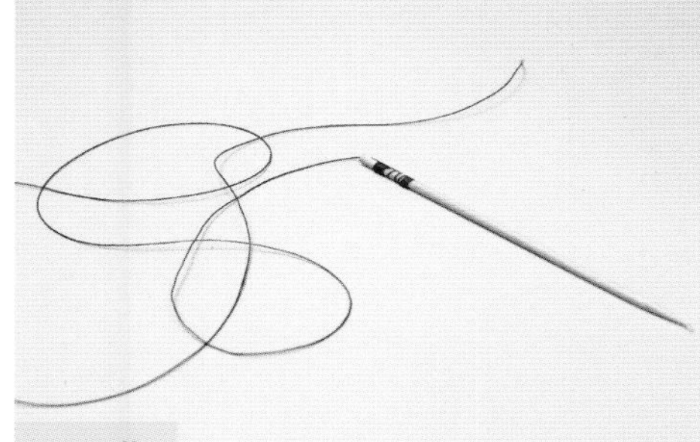
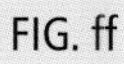
FIG. ff

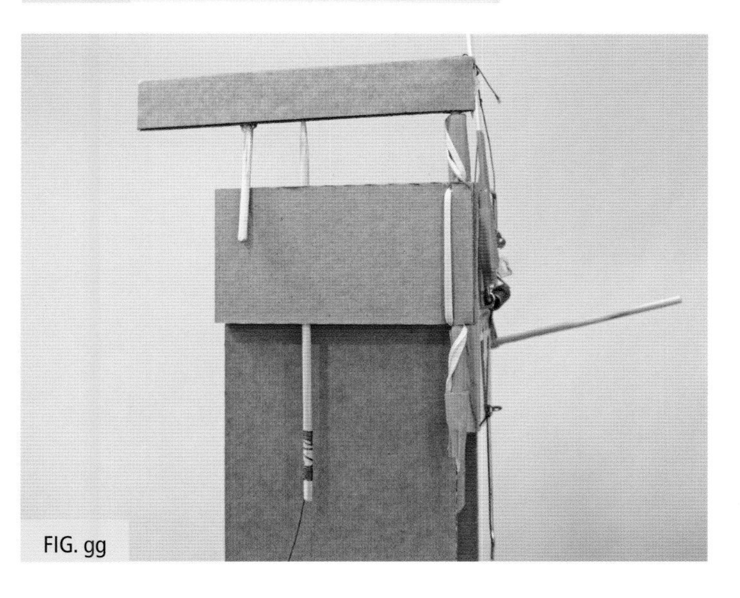
FIG. gg

### Attach the Pieces to the Mousetrap

1. This is where things could potentially get a little tricky. Mousetraps can be finicky beasts. You don't want your finger near one when it goes off. So, use good judgment and be careful. We need to attach the string to the craft stick, and we can do it one of two ways. The first way is to set the trap and then tie the string to the stick. I don't like this way because the trap could easily go off. The other way is to lower the trap but don't set it. Just keep the craft stick in your hand while you wrap the string around it. A little tape will help you hold it in place. Keep the skewer pin inserted while you do this step so you can leave a little slack. Once you get a feel for how much string you need between the craft stick and the skewer pin, then you can gently release the trap and give it a proper knot. Add a bit of glue and some tape over your knot. The craft stick is going to be pulling that pin really quickly, and if it's not attached well it could shoot off. I speak from experience **(fig. bb)**.

2. Swing over the lever attached to the rubber band until it's just over the mousetrap. Make a mark on the lever that's in line with the mousetrap catch **(fig. cc)**.

3. Rotate the lever one or two turns clockwise so there's some tension in the rubber band. Wind it enough so it wants to spring back into its resting position **(fig. dd)**.

4. Poke a hole in the underside of the lever right at the line you drew, and insert a piece of skewer. Make sure it's long enough to hit the catch when it closes **(fig. ee)**.

5. Tie a long piece of string to the dull end of the last skewer **(fig. ff)**.

6. Insert the skewer into the corrugation of the O piece and push up gently until you poke a small hole in the lever. The goal is to hold the lever in place until the pin is pulled **(fig. gg)**.

# START THE CHAIN REACTION!

### The Setup

1. Setting up the machine can be a bit tricky and finicky because a mousetrap is involved. This one in particular is slightly awkward. First, set the rubber band–loaded swing arm so it's out of the way. Now you can tackle the pin that will release the bottle. Lower the craft stick and insert the skewer pin into the first hole. You might have to raise the craft stick a bit to give it enough space to be inserted. You'll get a feel for how to do this after a bit of wiggling and testing. Once the pin is fully inserted and the craft stick is fully lowered, then you can set the hold-down bar on the mousetrap **(figs. a and b)**.
2. Carefully, so as to not set off the mousetrap, insert the soap bottle into its holder. Swivel the dispenser part to the back of the machine. Carefully add the water bottle into the holder. Start with about 8 ounces (235 ml) of water to see how much soap will squirt out at that weight. Then you can add or remove water accordingly **(fig. c)**.
3. Tape down the base of your machine to the table. Carefully tape the string from the skewer pin to the table 2" (5 cm) or so away from the base. You want some tension in the string, so it's best to hold the pin in place while you tape the string so you don't accidentally pull it **(fig. d)**.
4. Place the cantilever swing arm so the edge of it just comes into contact with the string.
5. Rotate the arm so it winds around the pencil structure, being careful not to bump the string and pull the pin. As you wind the arm, it will rise up higher and higher until there's no more string to wrap around the pencil **(figs. e and f)**.

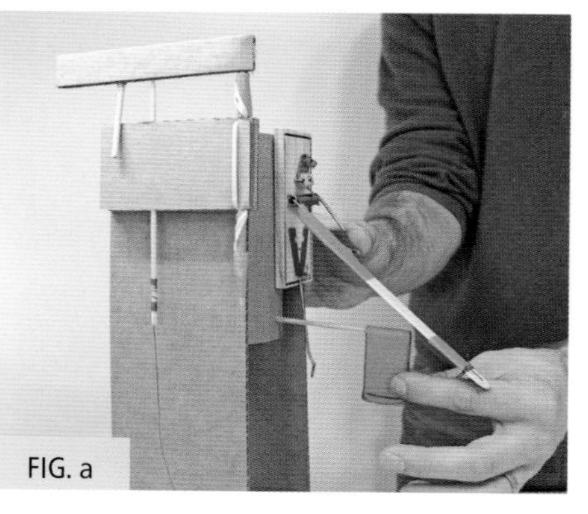
FIG. a

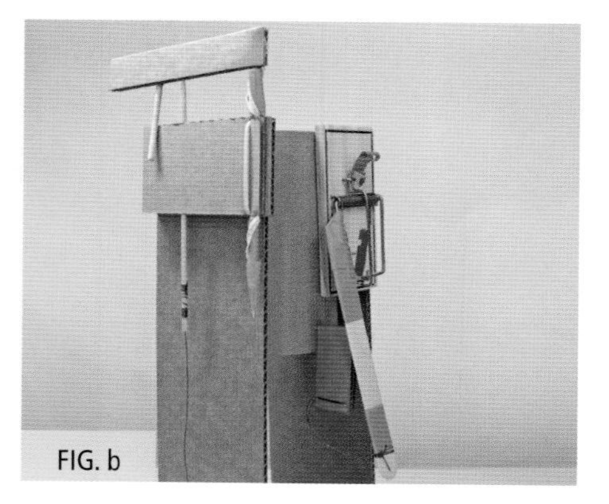
FIG. b

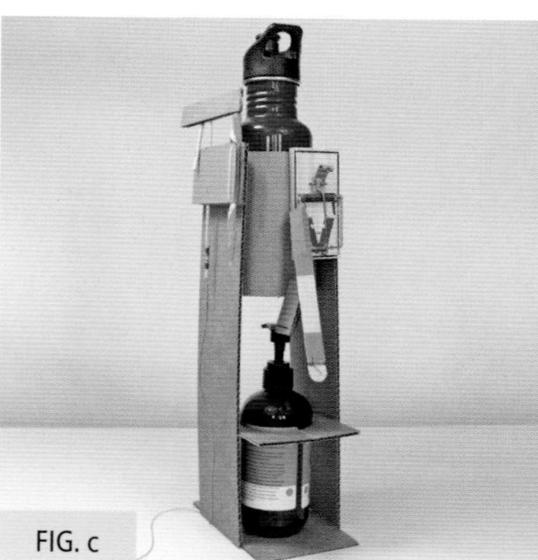
FIG. c

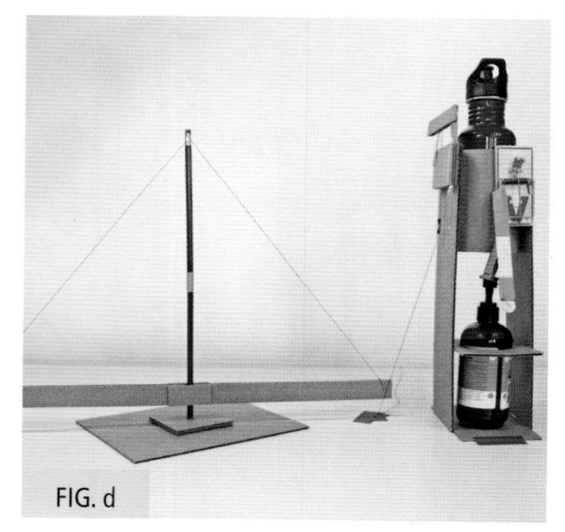
FIG. d

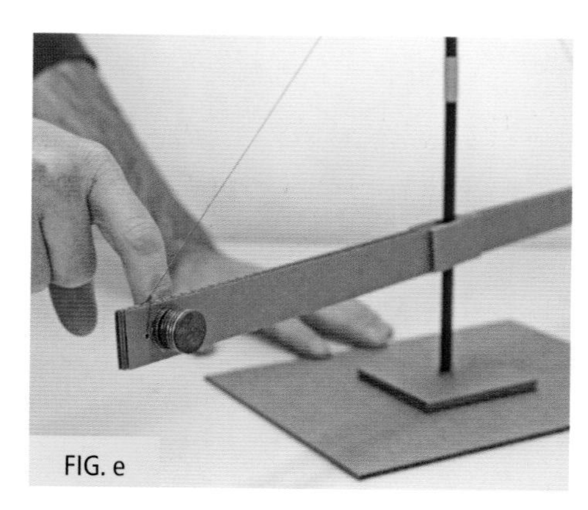
FIG. e

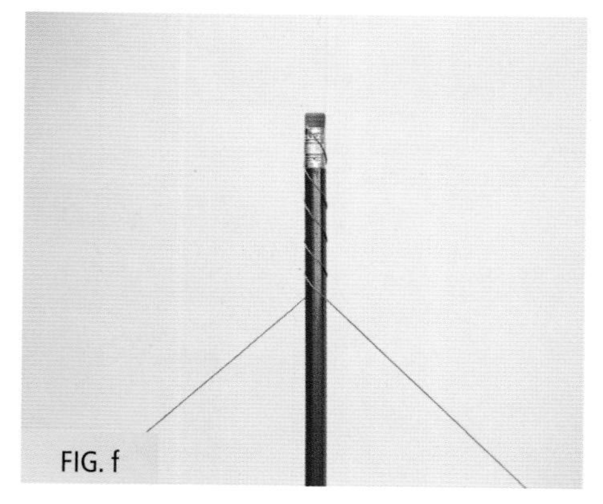
FIG. f

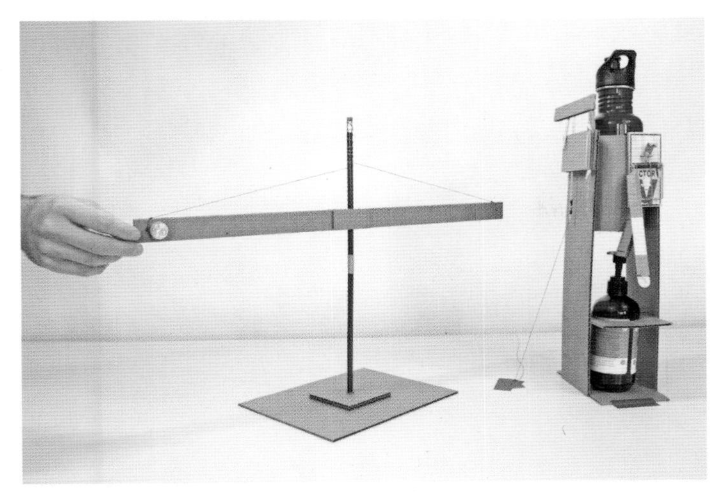

## Make It Go!

To start the chain reaction, simply let go of the cantilever arm. As it unwinds, it will get lower and lower, spinning the whole time. At the very bottom it will swing into the string, pulling out the pin. The lever attached to the rubber band will swing around, setting off the mousetrap. The mousetrap will pull the pin holding the water bottle. The water bottle will drop down, smashing into the soap dispenser, which will then dispense soap.

### Engineering Know-How

You can alter the amount of soap that's dispensed by changing one of two things. First, you could add more water to the water bottle, which would make it heavier. The heavier bottle would overcome the force of the spring inside the soap dispenser, causing it to bottom out and dispense a lot of soap. The second thing you can do is increase the distance between the water bottle and the soap dispenser. This increase in distance causes the water bottle to have more momentum.

FIG. g

FIG. h

### Troubleshooting

If you're having difficulty getting your mousetrap to close, you might want to look at the cardboard attached to your pin. If it's getting in the way, you can either trim the cardboard so it's smaller, or poke a new hole to the right side of the current one **(figs. g and h)**.

If you're rubber band arm doesn't want to swing, make sure it's attached like the one in the picture. If it looks correct, it might be the rubber band. Try to find a fresh, slightly thicker one **(fig. gg on page 79)**.

CHAPTER IIII

# MACHINES FOR FUN AND NONSENSE

Not all chain reaction machines serve a specific purpose. The four machines featured in this chapter aren't just fun to build, they're also sure to amuse!

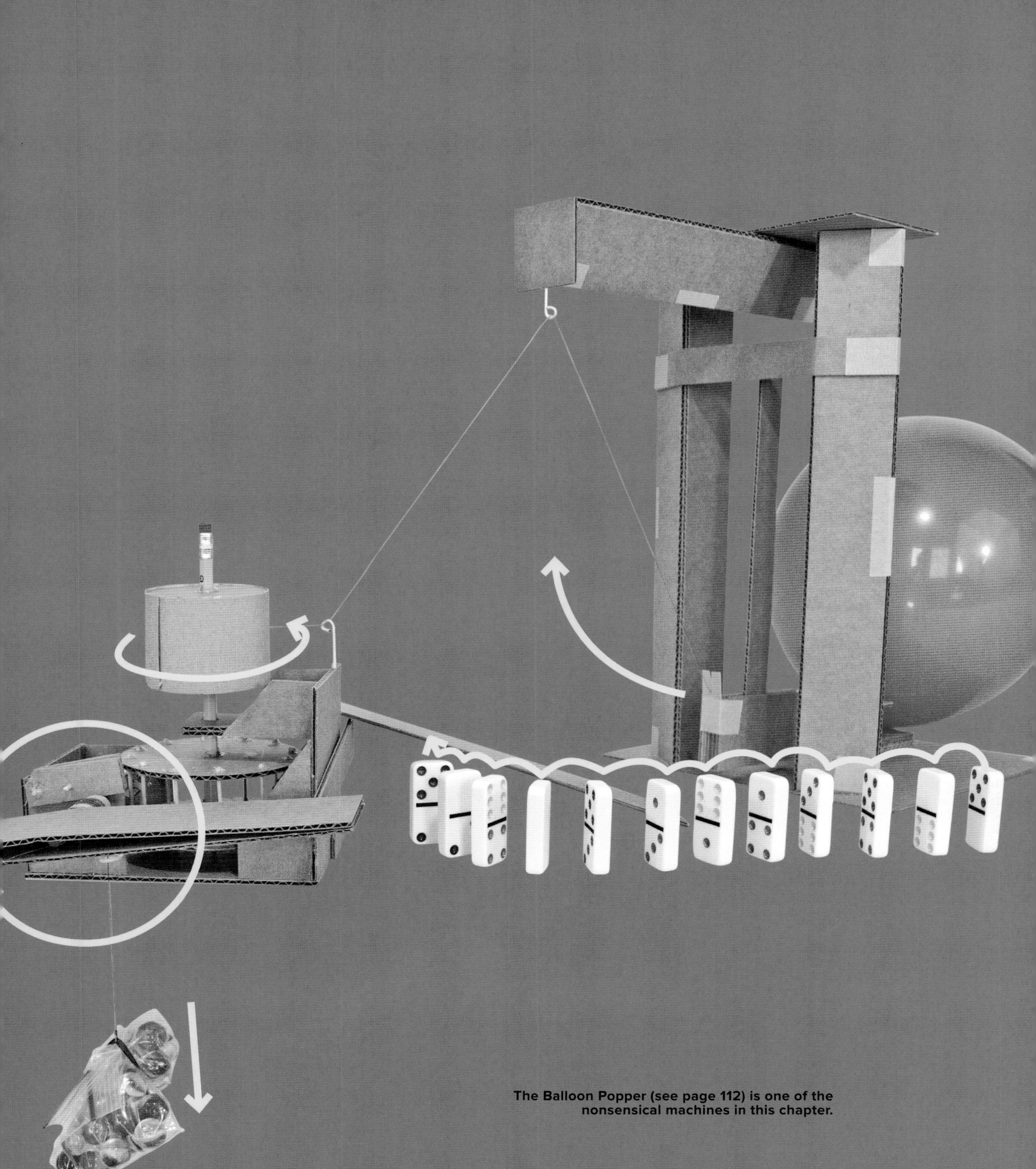

The Balloon Popper (see page 112) is one of the nonsensical machines in this chapter.

# FLAG RAISER

With just a few basic mechanisms and materials—a rack-and-pinion gear, some marbles, a ramp, and a weight—you can make a machine that can raise a flag!

# BUILD MATERIALS & TOOLS

## GEAR AND FLAG

**CARDBOARD:**

Two 2" × 12" (5.1 × 28 cm)–**A1 & A2**

1" × 12" (2.5 × 30.5 cm)–**B**

6" × 12" (15.2 × 30.5 cm)–**C**

Two circles 7" (17.8 cm) in diameter–**D1 & D2**

Two 1" × 12" (2.5 × 28 cm), with corrugation exposed–**E1 & E2**

2" × 3" (5.1 × 7.6 cm), with corner cut at an angle–**F**

2" × 3" (5.1 × 7.6 cm)–**G**

**CARDBOARD PIECES NOT SHOWN:**

2¼" × 12" (5.4 × 30.5 cm)–**H**

Two washers (see page 13)

**OTHER:**

Piece of ribbon or study but flexible string/thin rope–**I**

4" (10.2 cm) length of wire hanger or other stiff wire

Something light and colorful for your flag–**J**

Two unsharpened pencils (or 12" [30.5 cm] dowel)

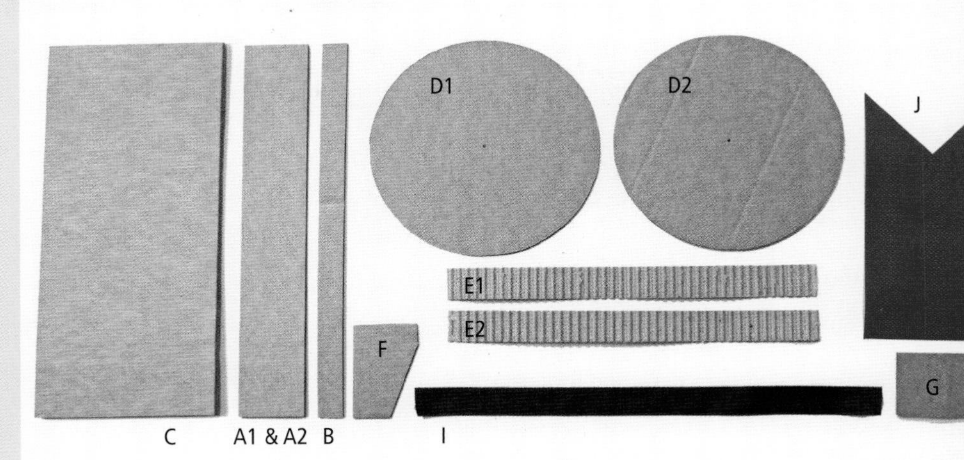


## MARBLE GATE STAND

**CARDBOARD:**

1" × 7" (2.5 × 17.8 cm)–**K**

Two 1" × 10¼" (2.5 × 26 cm)–**L1 & L2**

1" × 2" (2.5 × 5.1 cm)–**M**

1" × 2½" (2.5 × 6.4 cm), with corner cut out–**N**

1" × 4½" (2.5 × 11.4 cm), with triangular piece cut ¼" (6 mm) from the top–**O**

2" × 10" (5.1 × 25.4 cm), with sides scored ½" (13 mm) from the edge–**P**

2" × 3" (5.1 × 7.6 cm)–**Q**

2" × 2" (5.1 × 5.1 cm)–**R**

1½" × 2" (4 × 5.1 cm)–**S1 & S2**

1" × 4" (2.5 × 10 cm)–**T**

1½" × 8½" (3.8 × 21.6 cm)–**U**

2" × 5" (5.1 × 12.7 cm)–**V**

1½" × 2½" (3.8 × 6.4 cm)–**W**

1" × 1½" (2.5 × 3.8 cm)–**X**

3" × 5" (7.6 × 12.7 cm) with ½" (13 mm) square cutouts at the corners and all sides scored ½" (13 mm) from the edges–**Y**

**OTHER:**

Two toothpicks

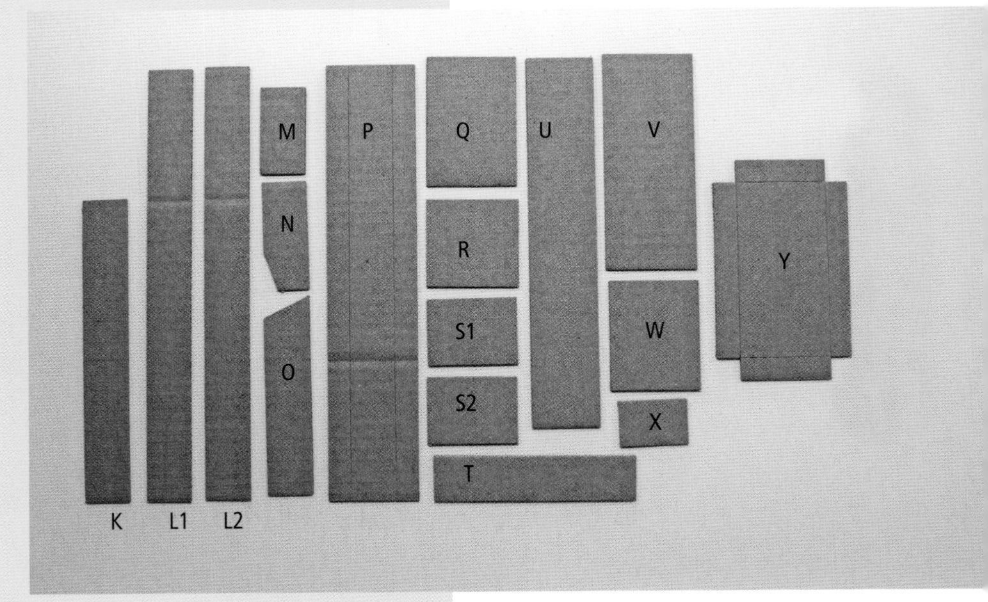


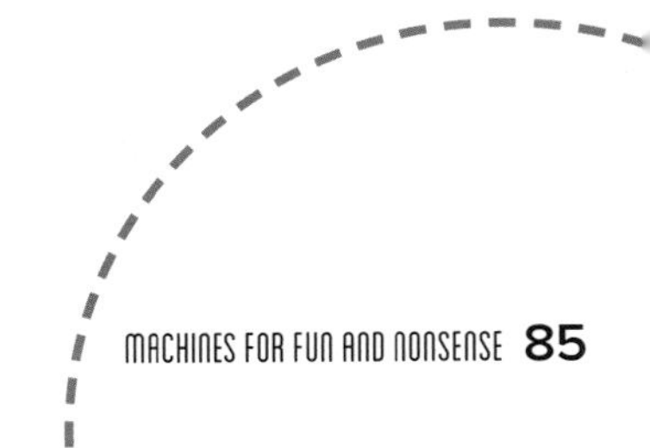

# BUILD THE MACHINE

## Make the Gear and Flag

1. Tear off the top layer of A1 to expose the corrugation. Glue A1 and A2 together with the exposed corrugation on top. This will be the rack for the rack-and-pinion gear **(fig. a)**.
2. Make a gear by wrapping the strips E1 and E2 around D1 and D2 (see page 18). We'll use this for the pinion in the rack-and-pinion gear. With an awl, make a hole in the center (see page 12) **(fig. b)**.
3. Attach C to the edge of the base (H). Add B to the front edge. This will allow the rack to slide back and forth **(fig. c)**.
4. Put the rack into the track you just made and place the pinion gear on top. Center the gear with the base. You want the rack and pinion to be touching but not be so tight it won't freely slide. Once you get it right, poke a hole through the wall of the base using the hole in the gear as a guide. Glue your ribbon or strong string (I) to the end of the rack **(fig. d)**.
5. Slide a piece of coat hanger through the hole and glue F with the corner cut off to the back for support. Make sure the coat hanger is level and not tilted in any direction **(fig. e)**.
6. Slide the pinion gear onto the coat hanger shaft and secure it with a paper washer (see page 17) **(fig. f)**.
7. Tape two pencils together to create a flagpole or use a 15" (38 cm) dowel. Glue or tape your flag (J) to your flagpole. Be mindful of the length and weight of your flag and pole. You want it to be as light as possible so it will be easier to raise **(fig. g)**.
8. Attach your flagpole to your gear. Be mindful of the position your rack is in when you glue on your flag. When the flag is vertical, the pinion should be flush on both ends. Attach G for support **(fig. h)**.

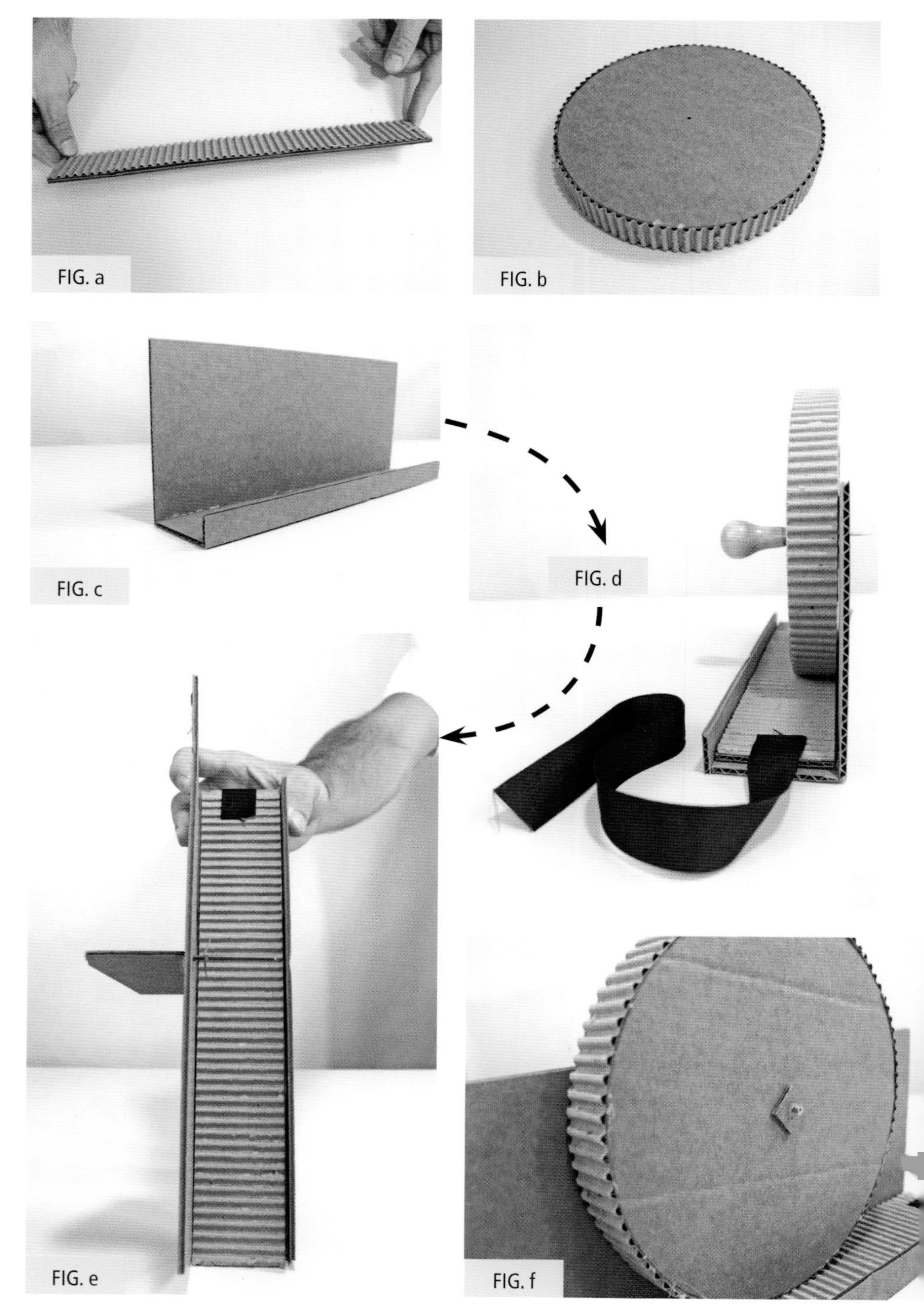

FIG. a

FIG. b

FIG. c

FIG. d

FIG. e

FIG. f

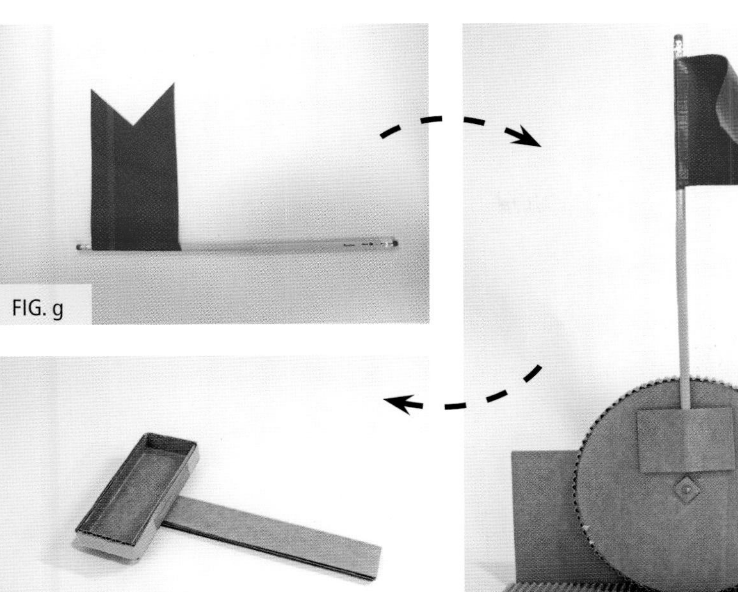

FIG. g

FIG. i

FIG. h

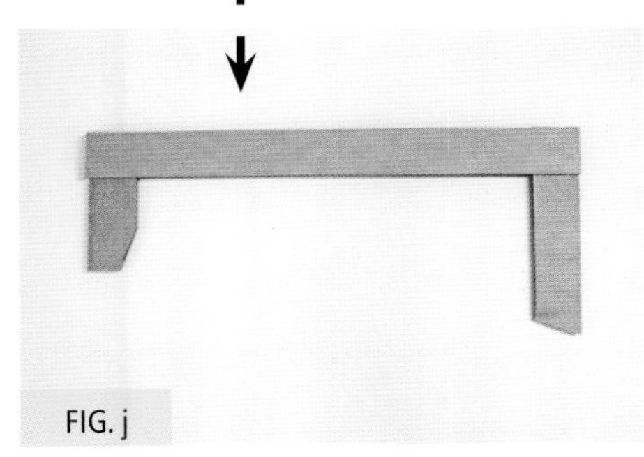

FIG. j

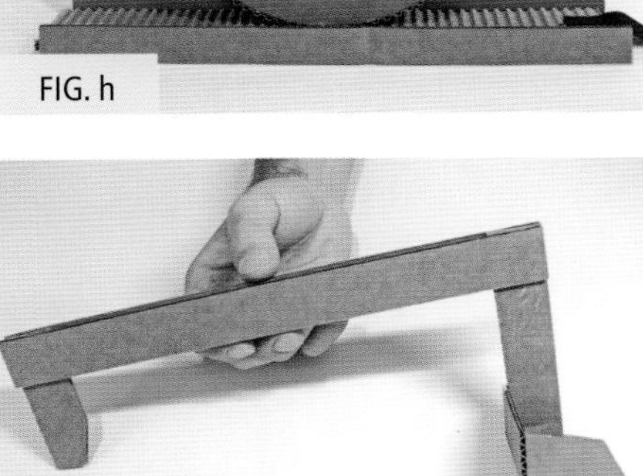

FIG. k

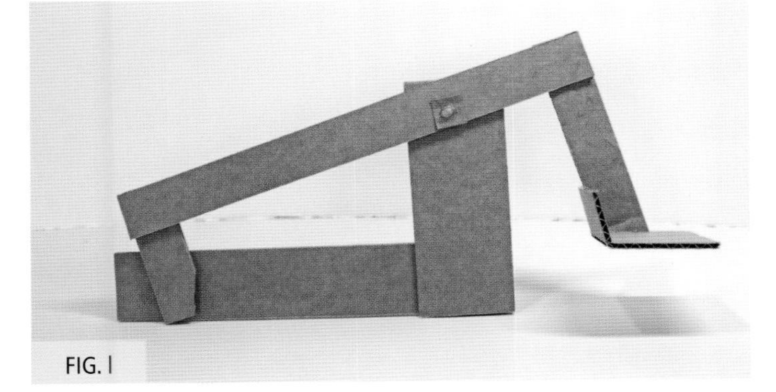

FIG. l

### Make the Marble Gate Stand

1. Fold up the edges of Y and tape the corners to create a shallow trough. Glue the box, centered, onto the end of K **(fig. i)**.
2. Attach N and O on either end of L1. Position the angles as shown in **fig. j**. Don't worry about your cuts being exact; we'll trim this later. Just eyeball it. Glue L2 on top of that, creating a sandwich. This piece will act as a marble gate later on.
3. Add R to O. Attach M for more rigidity **(fig. k)**.
4. Glue V to U, creating an "L" shape. Attach the marble gate to the base of the "L" shape 3" (7.6 cm) up the short end using a toothpick to create a hinge. Secure it with a paper washer (see page 17) **(fig. l)**.

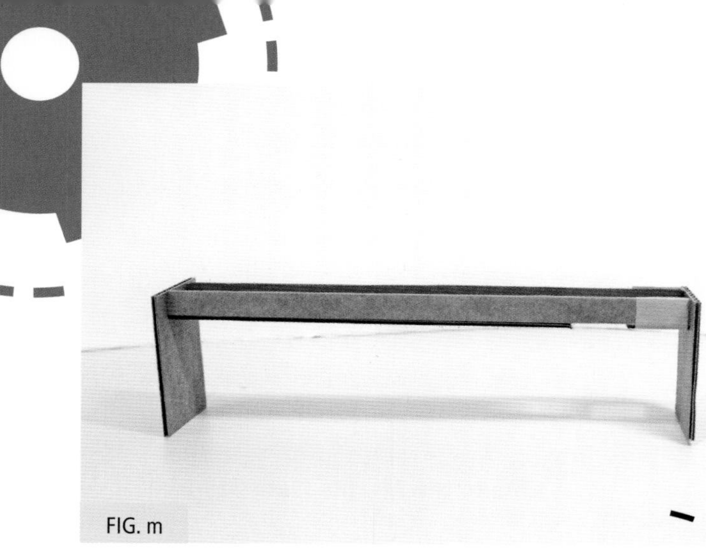

FIG. m

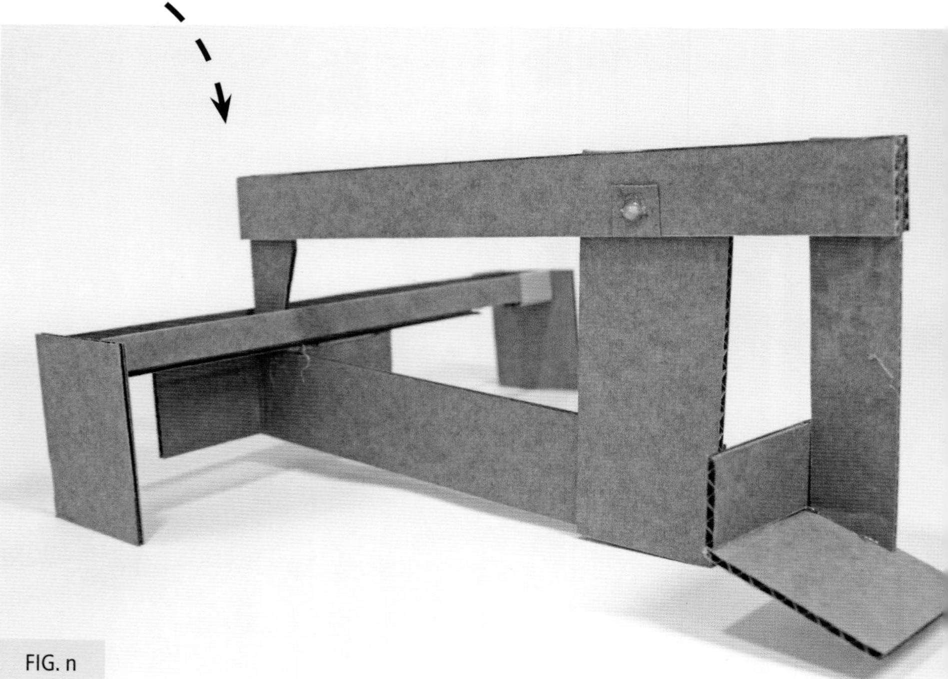

FIG. n

### Make the Marble Track

1. Fold P on the scored lines to create a channel. Glue Q to one end and W to the other. We want this to act as a very gradual ramp for our marbles to flow down **(fig. m)**.
2. Attach the marble gate to the underside of track 3½" (8.9 cm) from the taller end (Q). Attach S1 and S2 on either side of the bottom beam to add support. Now's the time to trim those end pieces. The one with the platform (R) should be okay, but you may need to take some scissors to the other side (N). When the platform R touches the ground, a marble should be able to pass under N. When the platform R is in its resting position, the other piece, N, should stop the flow of marbles. Carefully trim and adjust until it works for the marbles you're using **(fig. n)**.

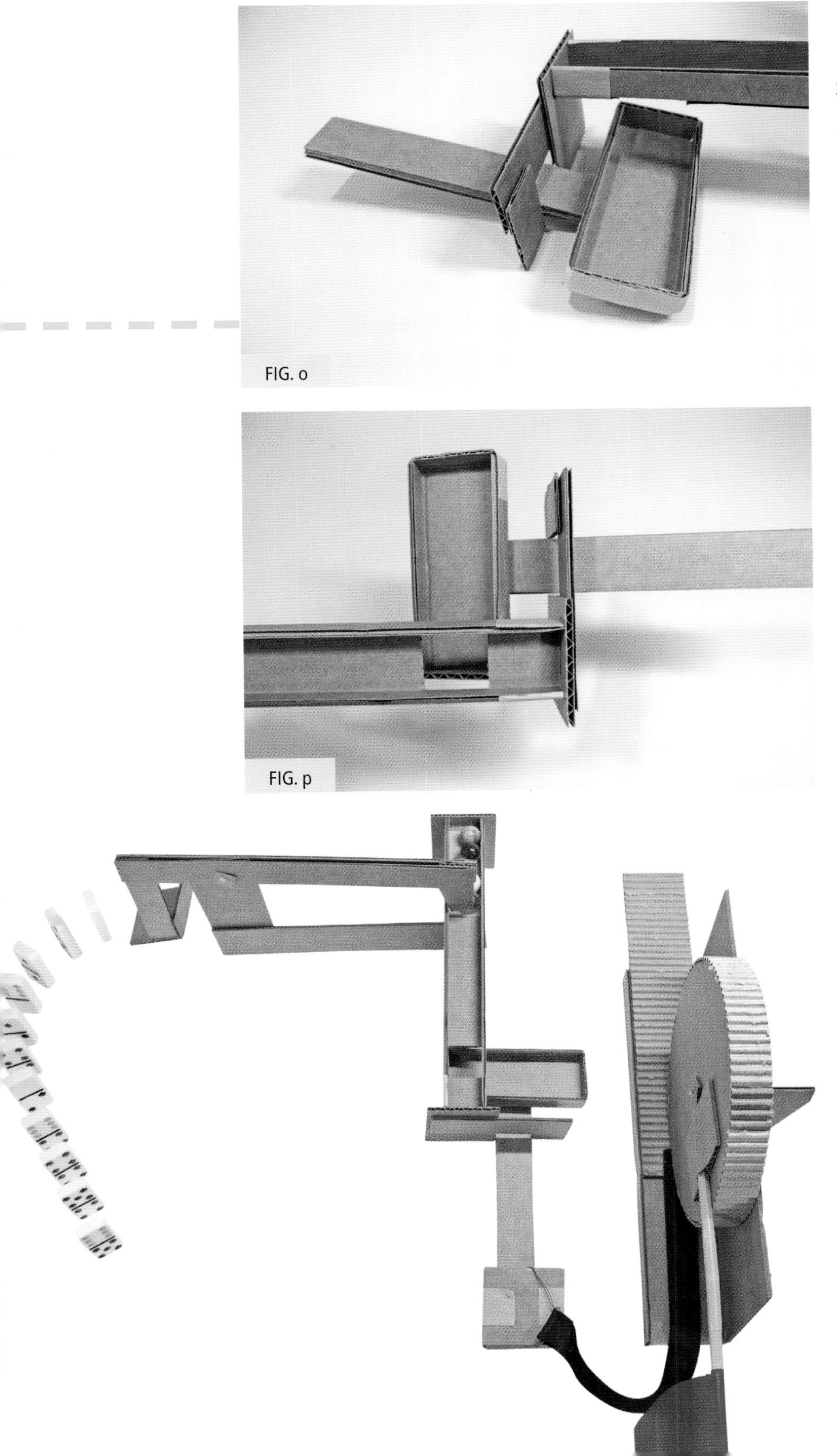
FIG. o

FIG. p

3. Now we'll attach the lever with the rectangular trough from fig. j. Stick a toothpick through the lever about 1" (2.5 cm) from the rectangular trough. Then poke that through the bottom support of the track (W), about ½" (1.3 cm) from the bottom. Add X for support to the other side of the toothpick. Poke this one slightly *less* than ½" (1.3 cm) from the bottom. That way the marbles will fall into the rectangular trough and then roll to the end. Otherwise, they might stay put and get stuck. Glue T to the top of X and the flat side of W to hold the end supports in place. Make sure there's enough play on either side of the lever between supports so the lever can easily go up and down **(fig. o)**.

4. With a utility knife, cut out a small hole in the track just above the rectangular trough. Make it as wide as the ramp and about 1" (2.5 cm) long. It needs to be big enough for the marbles to pass through without getting stuck **(fig. p)**.

5. We need a weight to add to the ribbon that's attached to our rack. It's best to use something with a hard, flat bottom, so if you don't have a small box, you can make one out of cardboard. A good size is 2" x 2" (5 x 5 cm) wide and 3" (7.6 cm) tall. I put batteries inside for the weights, but marbles or dried rice works fine, too. Tape the lid in place.

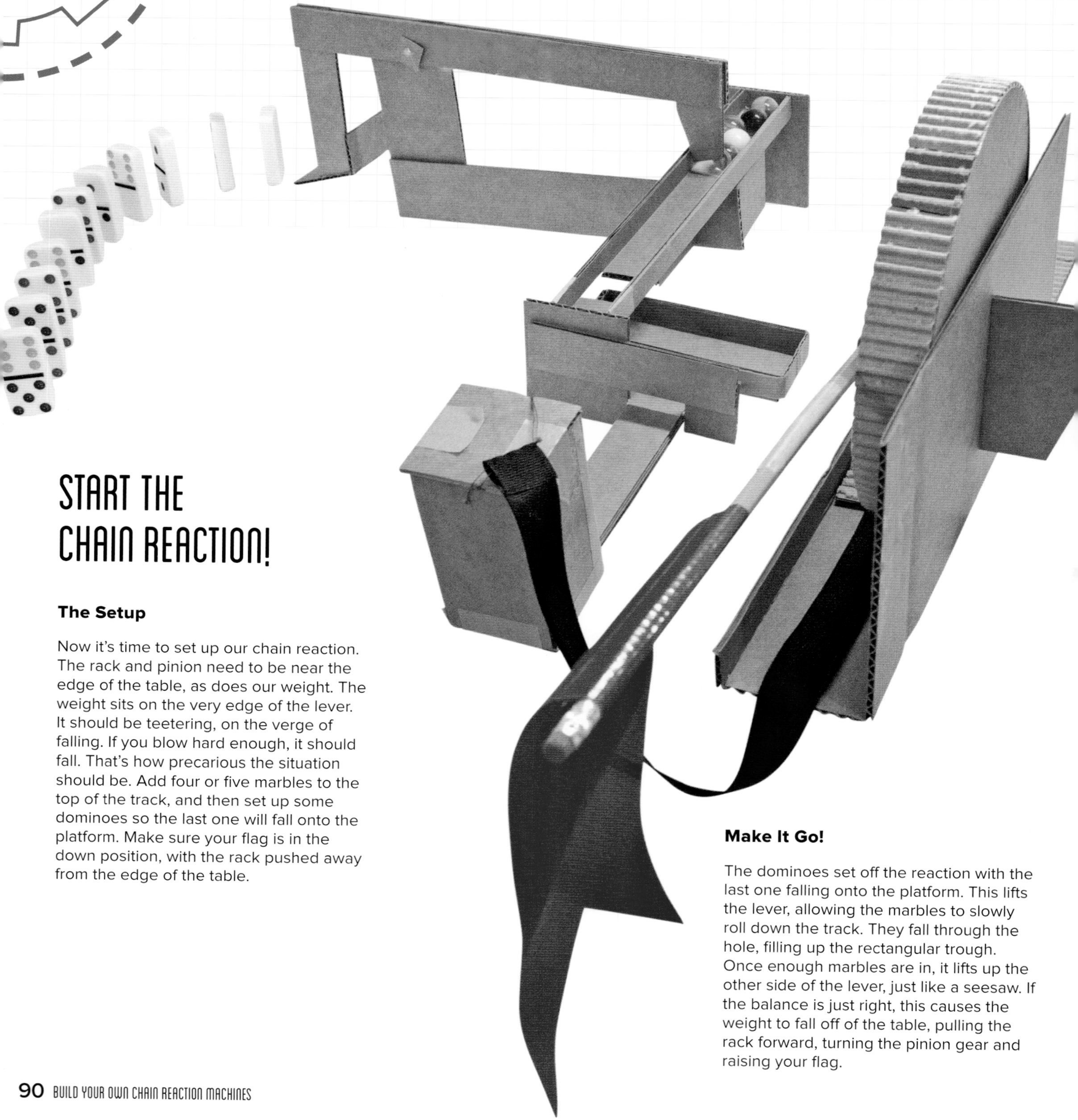

# START THE CHAIN REACTION!

### The Setup

Now it's time to set up our chain reaction. The rack and pinion need to be near the edge of the table, as does our weight. The weight sits on the very edge of the lever. It should be teetering, on the verge of falling. If you blow hard enough, it should fall. That's how precarious the situation should be. Add four or five marbles to the top of the track, and then set up some dominoes so the last one will fall onto the platform. Make sure your flag is in the down position, with the rack pushed away from the edge of the table.

### Make It Go!

The dominoes set off the reaction with the last one falling onto the platform. This lifts the lever, allowing the marbles to slowly roll down the track. They fall through the hole, filling up the rectangular trough. Once enough marbles are in, it lifts up the other side of the lever, just like a seesaw. If the balance is just right, this causes the weight to fall off of the table, pulling the rack forward, turning the pinion gear and raising your flag.

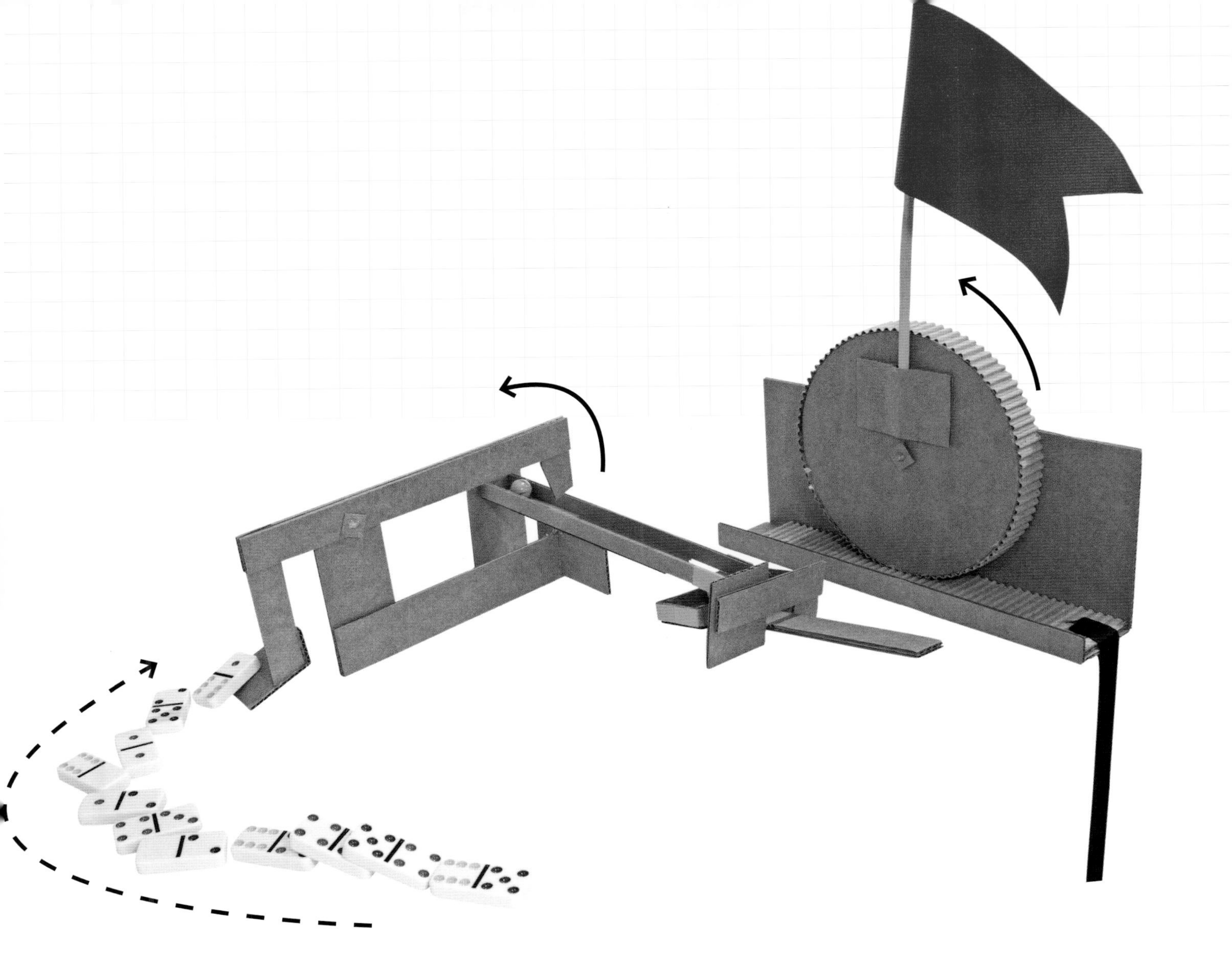

### Troubleshooting

It takes some patience and trial and error to get the weight to be balanced just right. Take your time with this and be methodical; add one marble (weight) at a time and test it. It makes it easier to get the balance to be spot-on.

Another issue you might run into is the rack and pinion gear mechanism. If you're having trouble, it's probably because the rack and pinion is either too loose or too tight. If it's too loose, the teeth won't engage or they'll only partially engage, which will wear them out pretty quickly. If they're too tight, the falling weight won't be able to slide the rack.

### Engineering Know-How

The released marbles cause the tipping over of the weight. Gravity causes the weight to fall, which pulls on the rack (straight gear). The movement of the rack causes the pinion gear to rotate 90 degrees. If the flagpole were longer or heavier, you would need a heavier weight or a bigger pinion gear to raise it.

# MARBLE LAUNCHER

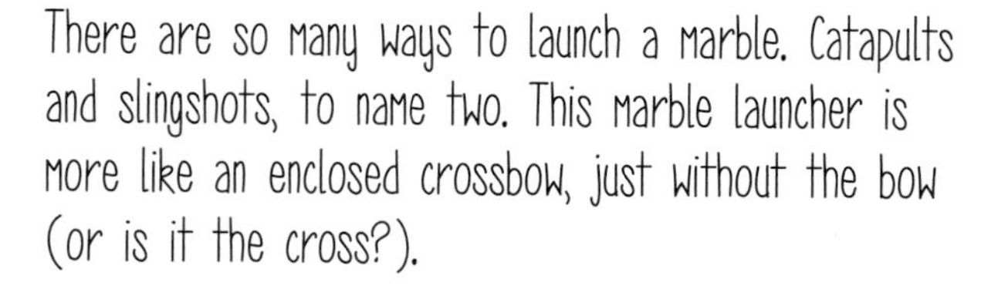

There are so many ways to launch a marble. Catapults and slingshots, to name two. This marble launcher is more like an enclosed crossbow, just without the bow (or is it the cross?).

# BUILD MATERIALS & TOOLS

## MARBLE LAUNCHING "TUBE"

**CARDBOARD:**

Four ½" x 1" (1.3 x 2.5 cm)–**A1, A2, A3 & A4**

Two ½" x 2" (1.3 x 5 cm)–**B1 & B2**

Eight ½" x 8" (1.3 x 20.3 cm)–**C1, C2, C3, C4, C5, C6, C7 & C8**

Two 2" x 8" (5 x 20.3 cm)–**D1 & D2**

**OTHER:**

6" (15.2 cm) craft stick–**E**

## MAIN BODY OF MARBLE LAUNCHER

**CARDBOARD:**

One 1½" x 1½" (3.8 x 3.8 cm)–**F**

One 6" x 9" (15.2 x 23 cm)–**G**

One 4½" x 11" (11.4 x 28 cm), scored lengthwise at 1½" (3.8 cm) and 3" (7.6 cm) (see page 14 for how to score)–**H**

Two 2" x 4½" (5 x 11.4 cm)–**I1 & I2**

Two 3" x 4½" (7.6 x 11.4 cm)–**J1 & J2**

3" x 6" (7.6 x 15.2 cm)–**K**

**OTHER:**

Toothpick or 2¾" (7 cm) piece of skewer or coat hanger–**L**

2" (61 cm) piece of string–**M**

Thin rubber band–**N**

*(continued)*

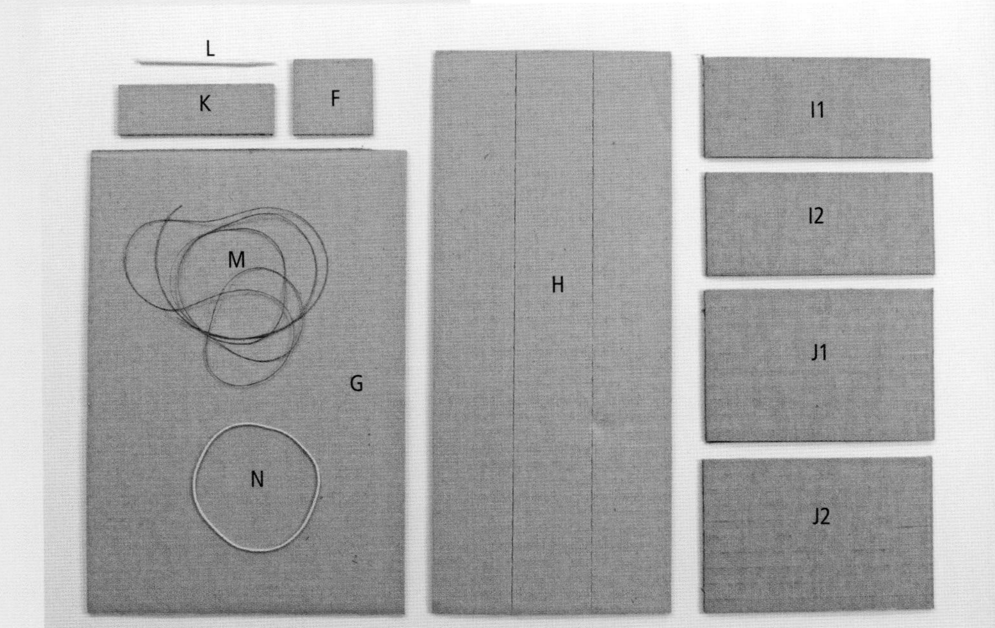

## MARBLE FUNNEL AND LEVER

**CARDBOARD:**

One 1" x 4½" (2.5 x 11.4 cm)–**O**

One 1" x 5" (2.5 x 12.7 cm)–**P**

Two 3" x 4" (7.6 x 10.2 cm), with 2" x 3" (5 x 7.6 cm) triangle cutout–**Q1 & Q2**

Two 1" x 9" (2.5 x 23 cm)–**R1 & R2**

**OTHER:**

A pencil, cut to 4½" (11.4 cm)–**S**

6" (15.2 cm) craft stick–**T**

Marbles

## PLATFORM FOR ZIGZAG RAMP

**CARDBOARD:**

One 1" x 3½" (2.5 x 9 cm)–**U**

One 1" x 4" (2.5 x 10.2 cm) thin cardboard, like a cereal box–**V**

Eight or nine 1" x 3" (2.5 x 7.6 cm)–**W1, W2, W3, W4, W5, W6, W7, W8 & W9**

One 5" x 13" (12.7 x 33 cm)–**X**

Two 1" x 13" (2.5 x 33 cm)–**Y1 & Y2**

## TOOLS

Hot glue gun and glue sticks

Awl

Painter's tape

Sharpened pencil

Ruler

Utility knife

Scissors

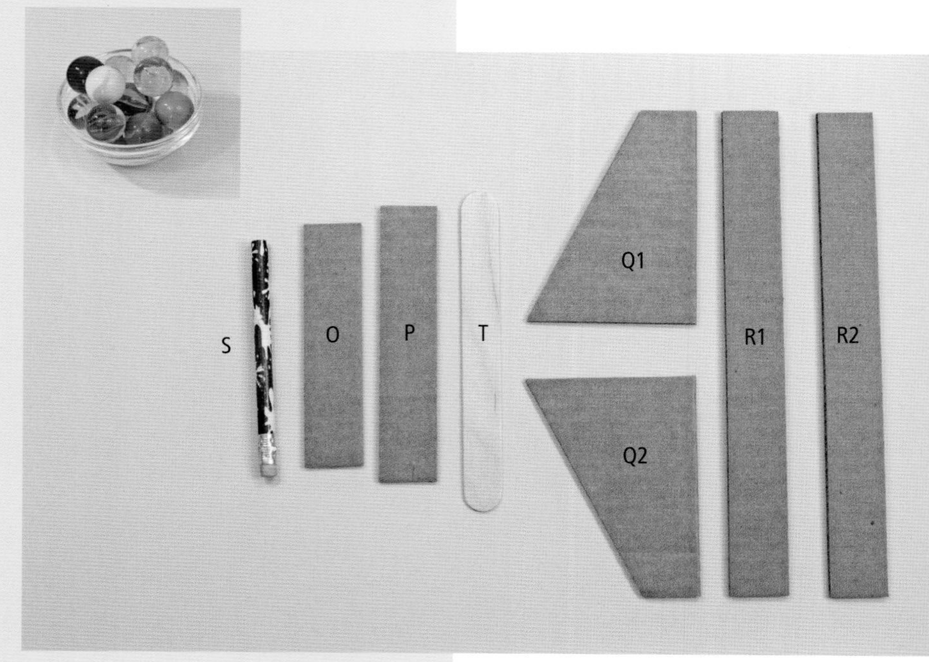

# BUILD THE MACHINE

### Make the Marble "Tube"

1. First up is the marble tube or track. This is what the marble will be launched out of. Lay out A1, A2, A3, A4, C1, C2, C3, C4, and D1 as shown in **fig. a**.

2. Start with the outer edges and work your way inward, gluing each piece as you go. The two short pieces act as spacers so we can fit a craft stick inside. That's what will push the marble out of the tube **(fig. b)**.

3. Glue C5, C6, C7, and C8 on top of the spacer **(fig. c)**.

4. Cap it off with D2 **(fig. d)**.

### Attach the Launching Mechanism

1. Cut the craft stick to 3" (7.6 cm) in length. Cut the rubber band so it's one long piece. Slide the craft stick into the gap in your marble launching tube. On each side, fold and glue B1 and B2. Make sure you don't get any glue on the tube. You want this piece to move freely up and down the tube, confined to the slit or gap in the track **(fig. e)**.

2. With an awl, poke hole in each piece of cardboard and attach the rubber band by sliding it through each hole and tying a knot. You want it tight enough so there's not a ton of slack when it's in a resting position **(fig. f)**.

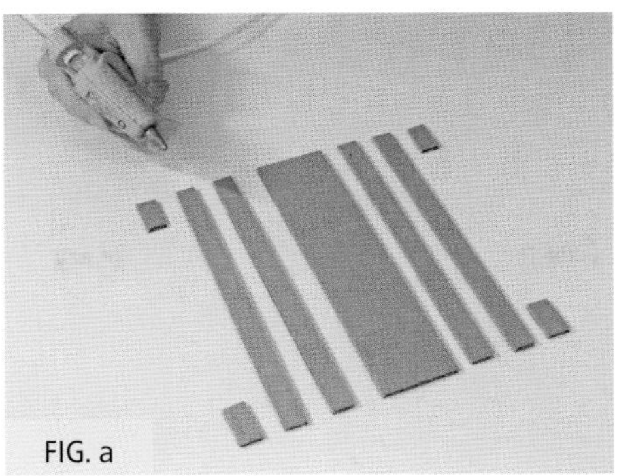

FIG. a

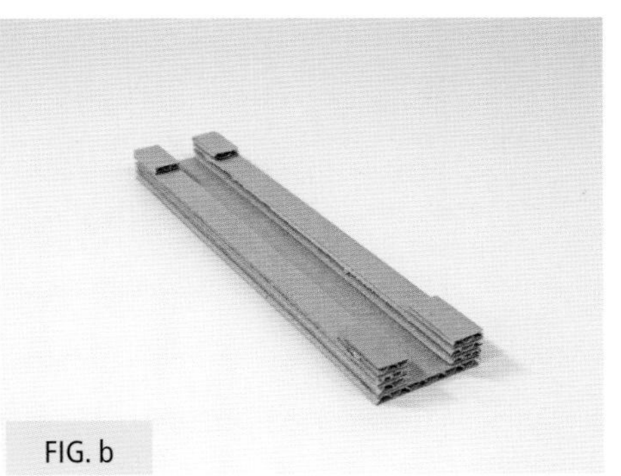

FIG. b

FIG. c

FIG. d

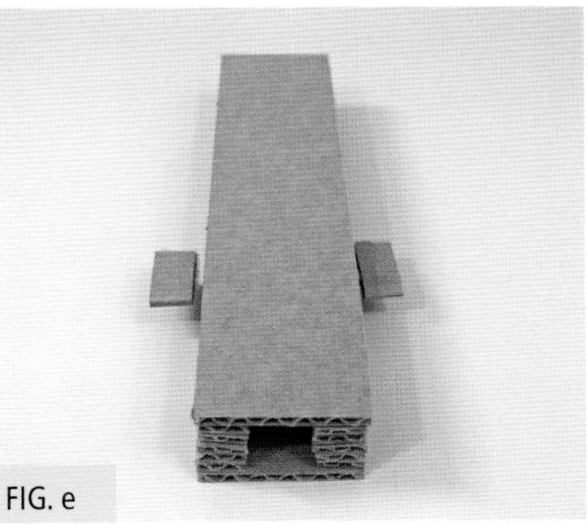

FIG. e

FIG. f

## Make the Main Body

1. Next up is making and attaching the tower to the base. Fold H on the scored lines to make a triangular tube. Use tape to hold it in place. Glue the triangular tube to the far corner of the base (G). Make sure the flat side is facing inward **(fig. g)**.
2. Glue together I1 and I2, and poke a hole in the center 1½" (3.8 cm) from the top. Using this piece as a guide, hold it next to your triangular tower and poke a hole at the same height **(fig. h)**.
3. Use a sharpened pencil to ream out the holes and make them larger. You want them big enough to allow your cut pencil to spin freely **(fig. i)**.
4. Sandwich the marble launching tube in between these two pieces. Make sure the end with the rubber band is positioned as shown. You want the launching tube angled upward so your marble will have a nice trajectory. The lower the angle, the farther it will go. The steeper the angle, the higher it will go. Use the sharpened pencil to keep everything aligned as you glue the marble tube to the triangular tower and the other piece to the tube and the base **(fig. j)**.
5. Remove the sharpened pencil and insert the cut pencil. Attach R1 to the pencil with a piece of tape. Hold the end of the strip while you rotate the pencil. Glue the strip as it slowly winds around the pencil **(fig. k)**.
6. Once this piece is all wrapped up, add R2 and keep winding. You're trying to create a cylinder around the pencil, at least 1" (2.5 cm) in diameter **(fig. l)**.

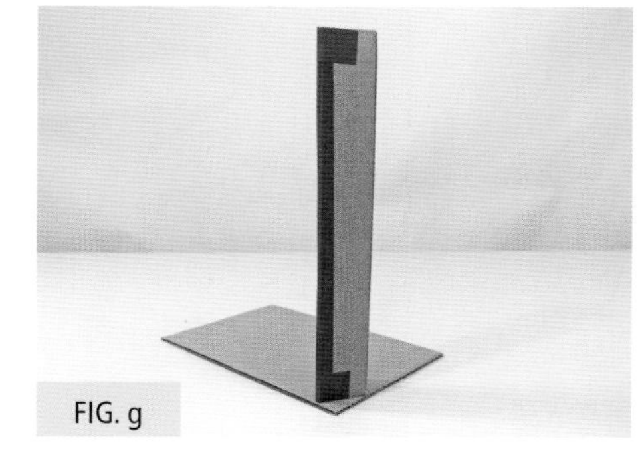
FIG. g

FIG. h

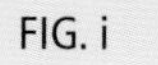
FIG. i

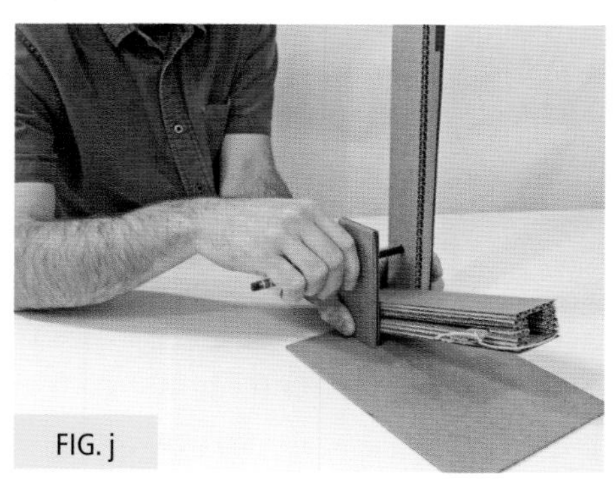
FIG. j

FIG. k

FIG. l

FIG. m

FIG. n

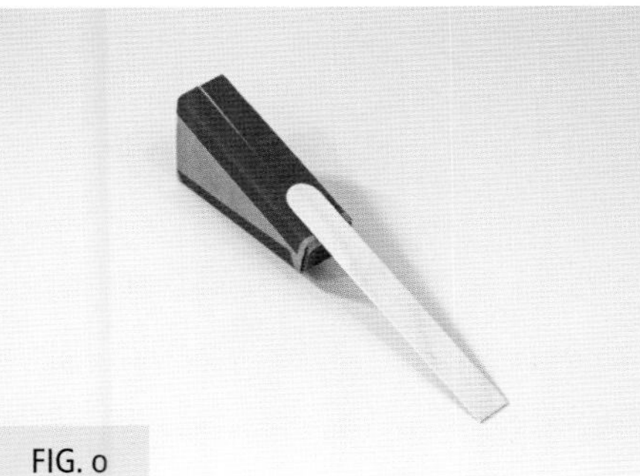
FIG. o

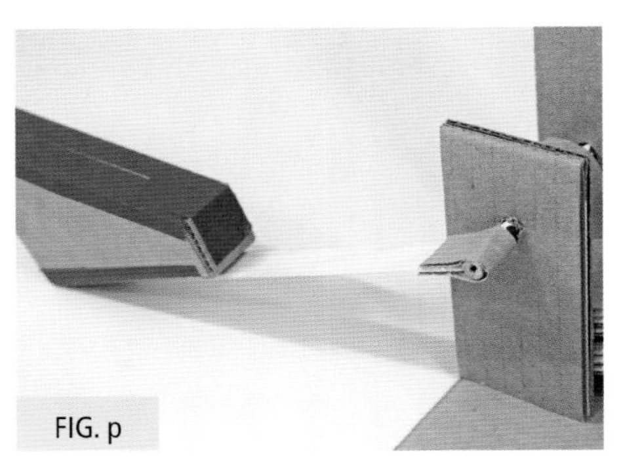
FIG. p

FIG. q

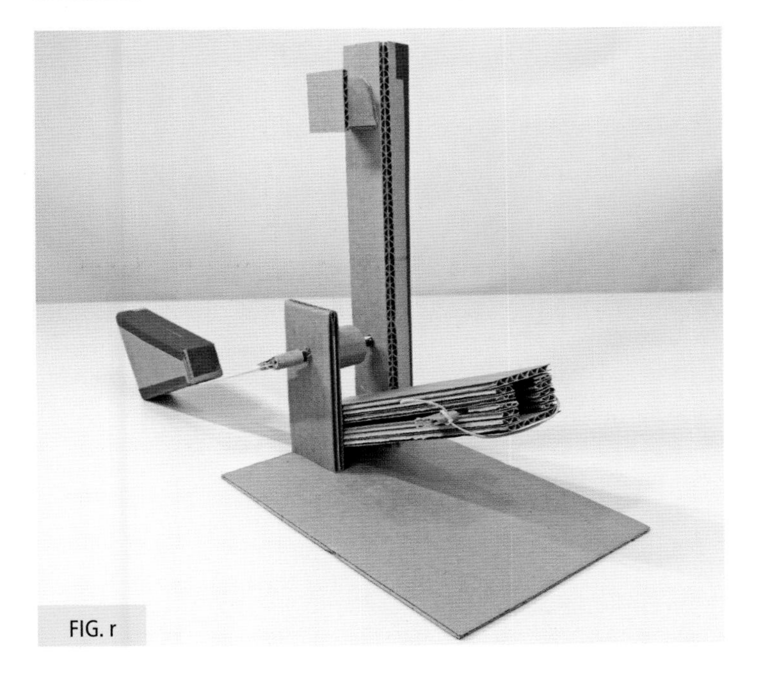
FIG. r

### Make the Marble Funnel

1. Lay out O, P, Q1, and Q2 for the marble funnel as shown in **fig. m**. You can tape them together when they're in this position and then fold everything together to create the funnel.

2. Tape the top flap down on top of the small opening. It doesn't have to look perfect. It just needs to be strong enough to hold some marbles and attach to a craft stick **(fig. n)**.

3. Cut the round section off one end of the craft stick and then glue it to the flat, non-slanted edge of your funnel. If you're using masking tape, the hot glue tends to not stick well to the slick tape. You can either go heavy on the glue or cover it in more tape after you glue it. Either should do the trick **(fig. o)**.

4. Glue the funnel lever to the pencil as shown. Make sure you don't get glue too close to the hole or it'll get jammed. You want this whole thing to rotate freely. For added structure you can fold and glue a scrap piece of cardboard around the connection between the craft stick and the pencil **(fig. p)**.

5. Shape N into a "U" shape and glue it to F **(fig. q)**.

6. Attach the U shape near the top of the tower, about 1" (2.5 cm) from the top. Make sure the U shape is upside down **(fig. r)**.

### Make the Zigzag Ramp

1. Glue Y1 and Y2 to either side of X. Cut out a half circle on the bottom right corner, approximately 1" (2.5 cm) wide **(fig. s)**.

2. Glue U to the bottom of the ramp, leaving the half-circle cut out open **(fig. t)**.

3. Take the thin piece of cereal box (V) and form it so it creates a circle with the half circle you cut from the ramp. Attach it with glue. Start from the bottom on the opposite side from the hole, and glue W1 at a slight angle. The goal here is to alternate W pieces to create a zigzag for the marbles to roll down. It'll slow them down and it looks cool, too **(fig. u)**.

4. Keep alternating the W pieces until you get to the top. The only rules here are to slant the pieces toward the middle and to make sure the spacing in the gap between each piece will easily let a marble through. If it turns out to be too tight, you can always trim it with scissors. No big deal **(fig. v)**.

5. Take J1 and J2 and cut them at an angle 1" (2.5 cm) from the top. Glue one of the angled pieces just above the upside-down U. Make sure the tilt is facing away from the marble launching tube. Glue the other angled piece on the opposite side in the same spot. Use a ruler to make sure both pieces are even. This is pretty important. If you're pieces are off by too much, the zigzag ramp will tilt too much to one side, and your marbles might get stuck. You can always trim or shim it, but if you're patient it's easy enough to get right the first time **(fig. w)**.

6. Hold up your funnel lever so it's at a slight angle, as shown in the picture. Place the zigzag ramp on top, and try to line up the hole in the ramp with the funnel. Attach it with hot glue and double-check the position before the glue cools down **(fig. x)**.

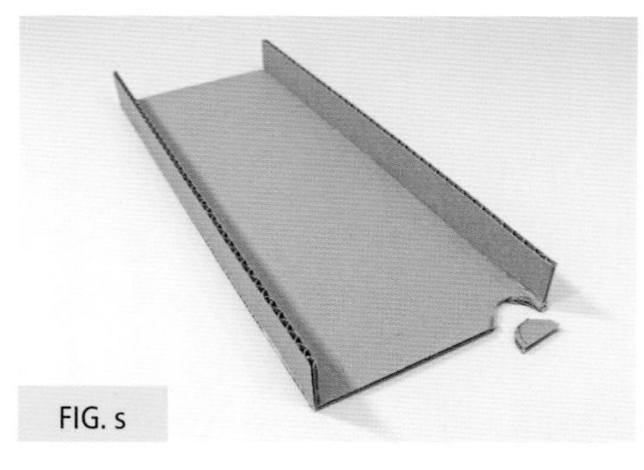

FIG. s

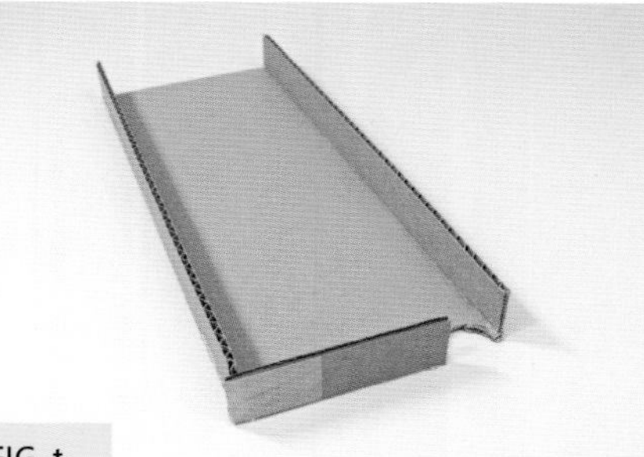

FIG. t

FIG. u

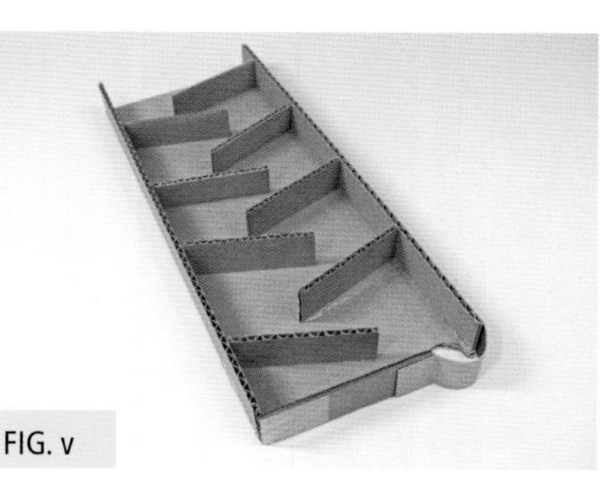

FIG. v

FIG. w

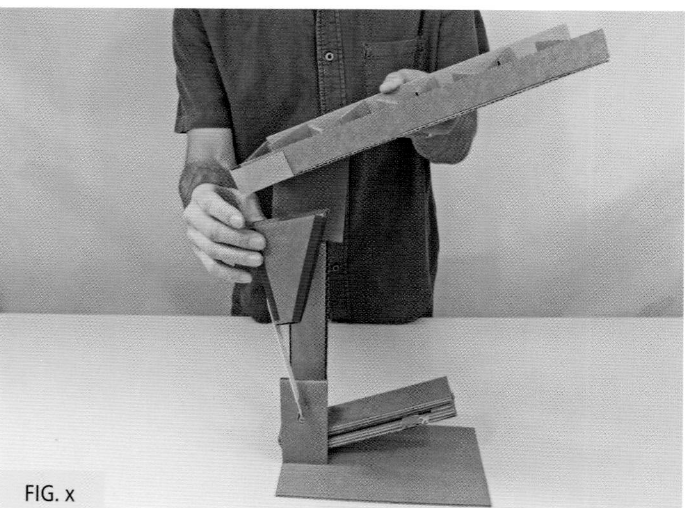

FIG. x

FIG. y

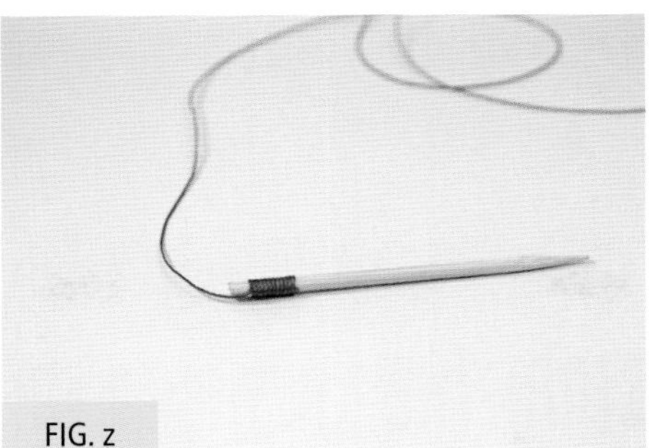

FIG. z

FIG. aa

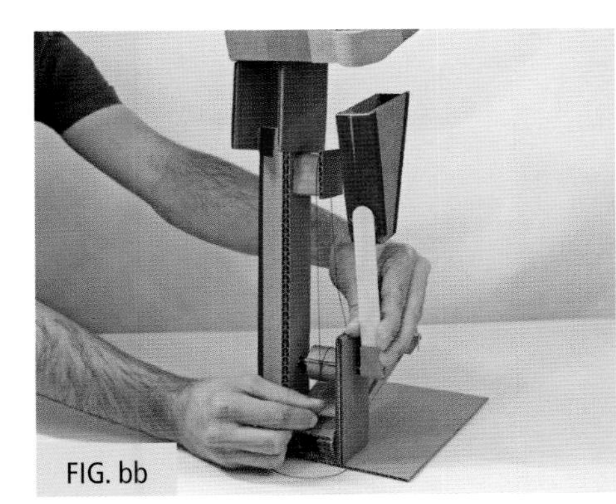

FIG. bb

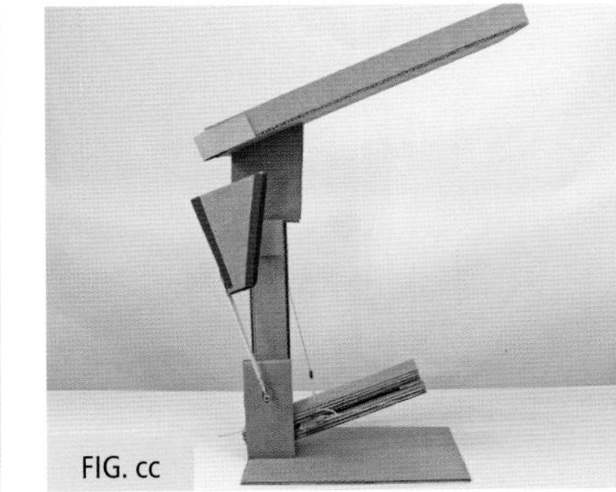

FIG. cc

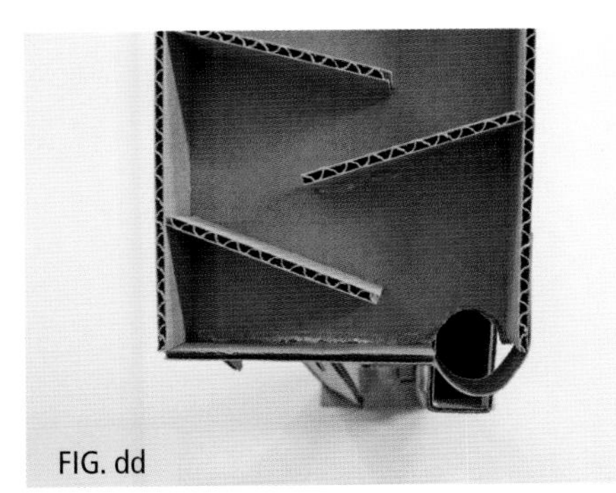

FIG. dd

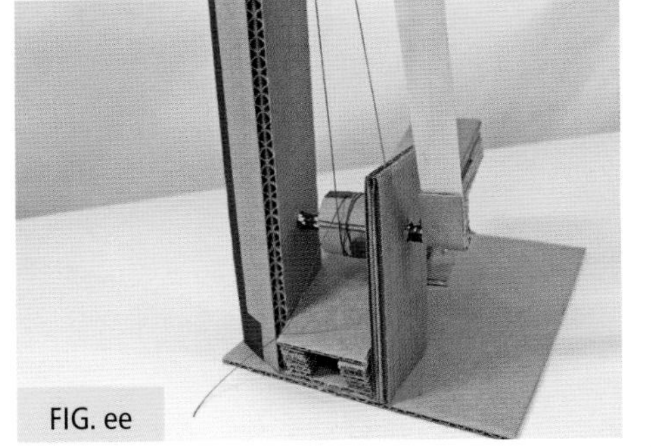

FIG. ee

### Make the Trip Wire

1. Pull back on the marble launching mechanism as far as it will go. Use an awl to poke a hole just in front of the edge of the craft stick. Go through to the bottom section of the launching tube. In a second we'll enlarge this hole slightly to fit your toothpick, skewer, or wire **(fig. y)**.

2. Tie a 2' (61 cm) piece of string around the end of your toothpick, skewer, or wire **(fig. z)**.

3. Insert the toothpick into the hole. It should pull out easily. If it doesn't, carefully wiggle it around to enlarge the hole. Place the string over the U-shaped piece and bring it down to the other side. **(fig. aa)**

4. Hold up the funnel so it's in line with the hole in the zigzag ramp. You might need to grab another person to help you out with this part. With the funnel lined up and the skewer pushed all the way inside the hole of the marble tube, wrap the end of the string around the cylinder on the pencil, and place a piece of tape over it to hold it in place **(fig. bb)**.

5. This is how it should look from the side and top **(figs. cc and dd)**.

6. Double-check to make sure everything is in the right place. (The string has a tendency to slip through the tape). Add more tape and even some hot glue to make sure the string is totally secure **(fig. ee)**.

# START THE CHAIN REACTION!

### The Setup

Slide back the launching mechanism and insert the pin to hold it in place. Insert a marble into the tube.

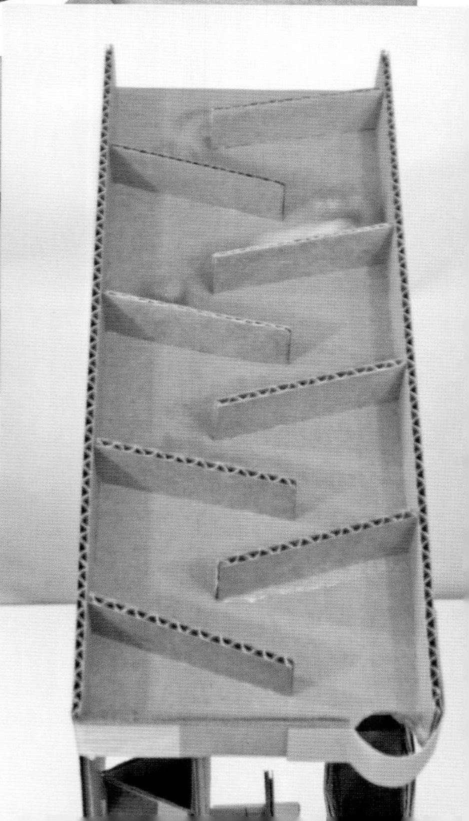

### Make It Go!

Place a handful of marbles on the top of the zigzag ramp. As they roll down to the bottom, they'll fall into the funnel. Once enough marbles fall, the funnel will be heavy enough to fall over, pulling out the pin. Once the pin is released, the marble in the tube will be launched!

## Troubleshooting

If the rubber band gets in the way, you can add a piece of tape to hold it down. Just make sure it doesn't lift up on the inside; the stickiness will get in the way of the marble **(fig. a)**.

If your marbles get stuck on one side of your zigzag ramp, it's probably not balanced. Mine wasn't perfectly level, but it still worked out okay. If it gets too out of whack, you can add a shim under the entire base of the machine to level it out **(fig. b)**.

If your marbles fall onto the table instead of going into the funnel, there are a couple of things you can do. The first is to just squeeze the funnel so it's wider **(fig. c)**. Or you can build a new funnel that's larger in all dimensions. You can also attach a short tube where the marbles come out of the zigzag ramp. This will help guide the marbles into the ramp. Just make sure it's short enough so it doesn't interfere with the funnel when it falls.

If the funnel falls over but doesn't pull the pin, you can try to adjust everything, but it still might not work. A fix could be to undo the string and make the cylinder larger by adding another strip of cardboard **(fig. d)**. You'll need more marbles (more weight) to lower the ramp, but the larger cylinder will pull more of the string, so you have a better chance of clearing the hole, releasing the launching mechanism.

FIG. a

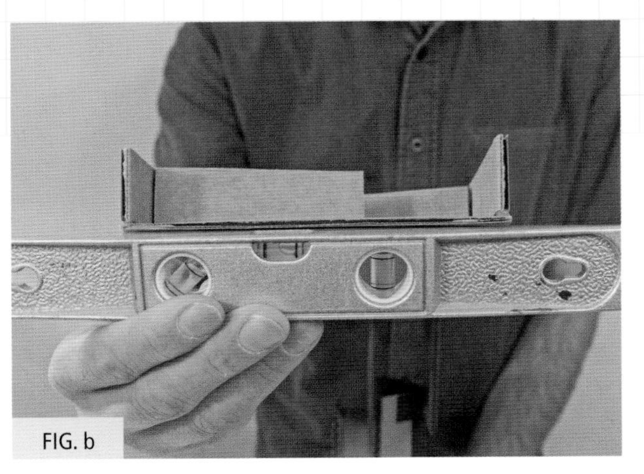

FIG. b

FIG. c

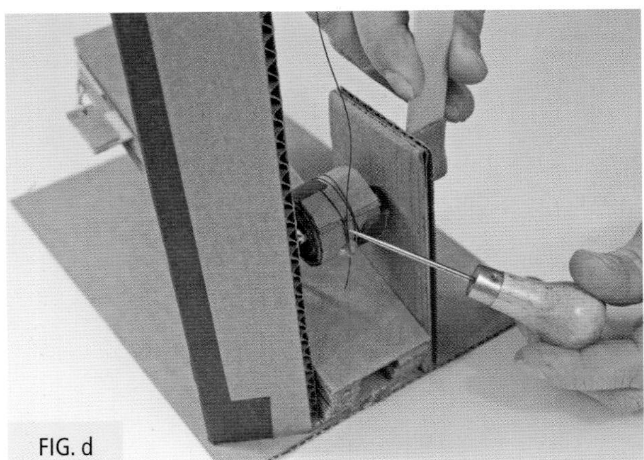

FIG. d

### Engineering Know-How

The angle of your launching tube can change the trajectory (path in the air) of your marble. If you create a very steep launch angle, your marble won't go very far, but it should go quite high. If you put your launching tube essentially flat, your marble won't go into the air at all, but will go much farther.

# MUSIC MAKER

"Music" is a generous definition for the sounds this machine makes. It's like a cross between a xylophone and someone dropping a drawer full of silverware. It's still a load of fun, though.

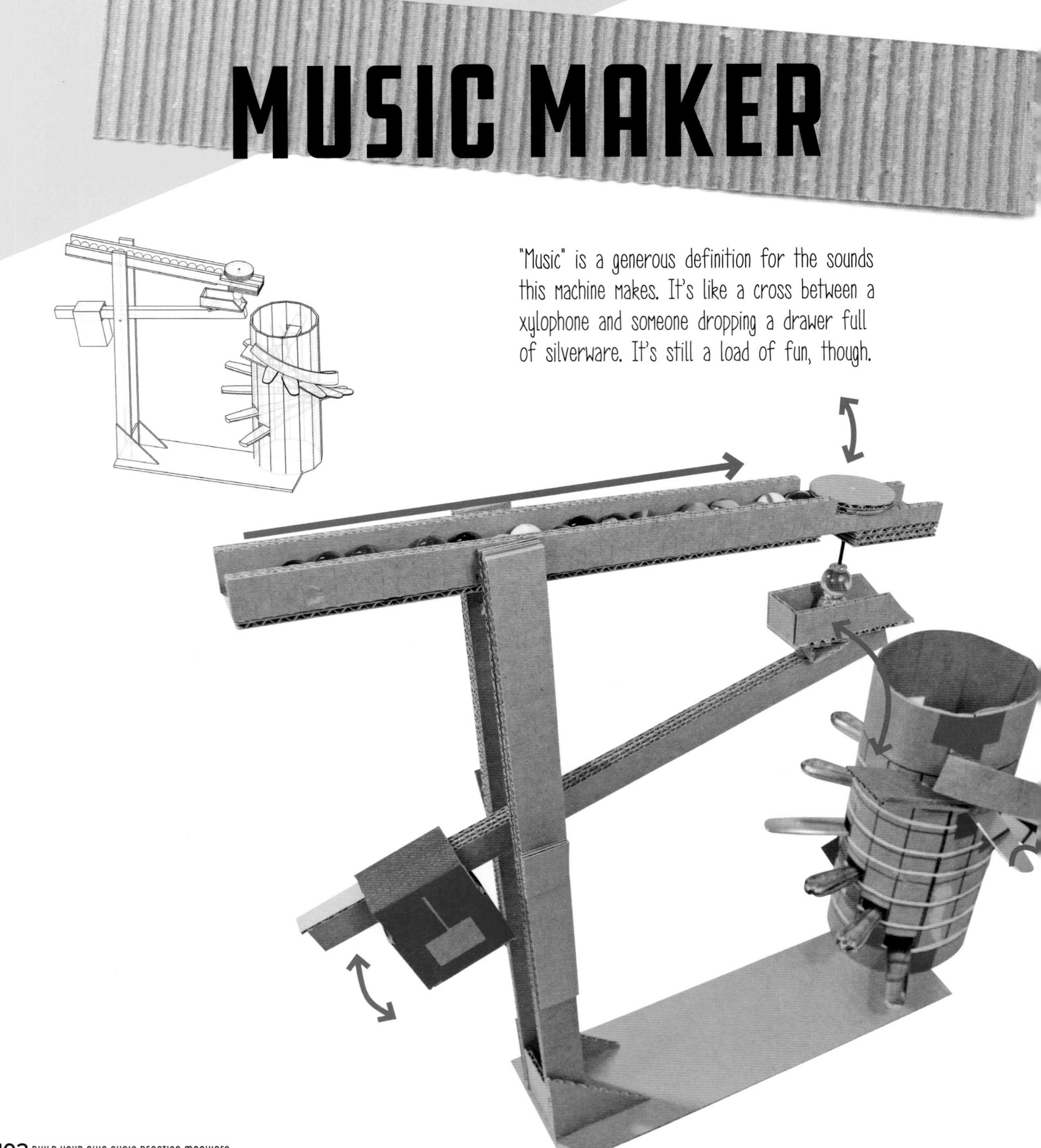

# BUILD MATERIALS & TOOLS

## MUSIC MAKER

### CARDBOARD:

One 2" x 3" (5 x 7.6 cm)–**A**

Four 1" x 1" (2.5 x 2.5 cm)–**B1, B2, B3 & B4**

One 1" x 16" (2.5 x 40.6 cm)–**C**

One 11" x 16" (28 x 40.6 cm)–**D**

### OTHER:

14 rubber bands–**E**

7 butter knives (It's actually better if you have an assortment of knives. You'll get more interesting sounds.)–**F**

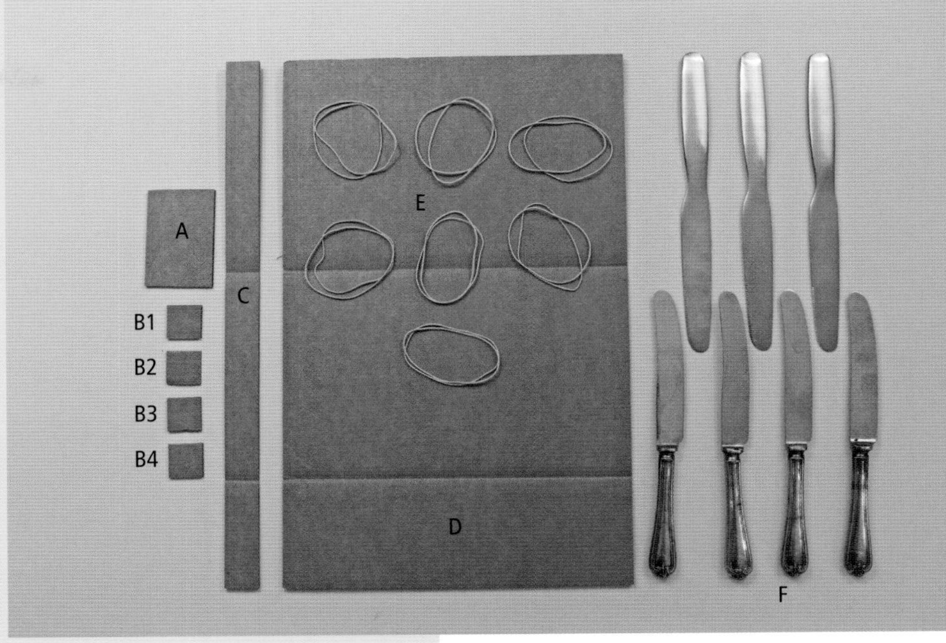


## MARBLE MACHINE

### CARDBOARD:

Four 1" x 18" (2.5 x 45.7 cm)–**G1, G2, G3 & G4**

One 3" x 3" (7.6 x 7.6 cm), cut on the diagonal into two triangles–**H**

Two 1" x 12" (2.5 x 30.5 cm)–**I1 & I2**

Five 1½" x 2" (3.8 x 5 cm)–**J1, J2, J3, J4 & J5**

Two 1" x 1½" (2.5 x 3.8 cm)–**K1 & K2**

Two circles, 2" (5 cm) diameter–**L1 & L2**

One 4" x 13" (10.2 x 33 cm)–**M**

Four ¾" x 14" (1.8 x 35.6 cm)–**N1, N2, N3 & N4**

One ¾" x 1" (1.8 x 2.5 cm)–**O**

Two ¾" x 3" (1.8 x 7.6 cm)–**P1 & P2**

One 1¼" x 3" (3.1 x 7.6 cm)–**Q**

One 1½" x 10½" (3.8 x 26.7 cm)–**R**

Four ¾" x 14" (1.8 x 35.6 cm) with a ⅜" x 3" (1 x 7.6 cm) cutout at one end–**S1, S2, S3 & S4**

Two 1" x 2" (2.5 x 5 cm)–**T1 & T2**

Two 1¼" x 2" x (3.1 x 5 cm)–**U1 & U2**

Two 2" x 3" (5 x 7.6 cm)–**V1 & V2**

### OTHER:

2" (5 cm) piece of wire–**W**

Marbles–**X**

Toothpick–**Y**

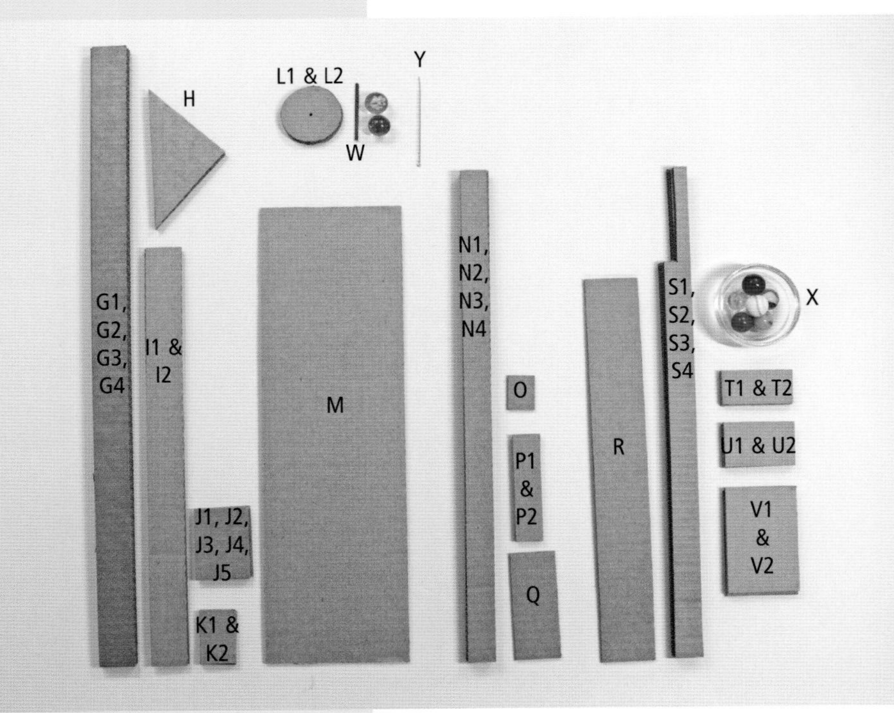


## TOOLS

Pencil

Ruler

Utility knife

Hot glue gun and glue sticks

Painter's tape

Awl

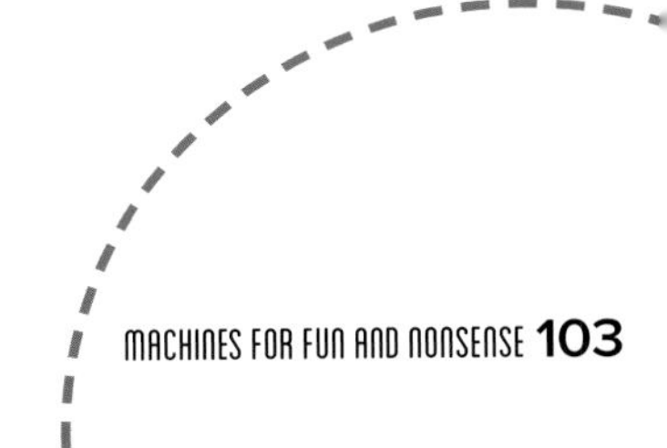

# BUILD THE MACHINE

### Make the Music Maker

1. Starting with D, draw a line 2" (5 cm) down from the top edge of the long side. Then draw a line every 1" (2.5 cm) below that until you have 7 rows. Next, draw a line 1" (2.5 cm) from the left side. Keep making lines until you have 14 columns. Starting with the top left box, shade a rectangle within the box that's ½" (1.3 cm) high. It helps to shade it so you know which pieces you're going to cut out later. Keep doing this in a diagonal line until you get to the bottom row. Then go up to the top again, move over 7 boxes, and do the same thing, except instead of a ½" (1.3 cm) tall rectangle within the box, make one that's ¾" (1.8 cm) tall. Keep doing this in a diagonal line until you get to the bottom row. With a utility knife, carefully cut out all of the shaded boxes **(fig. a)**.

2. With a utility knife, score the cardboard along the line that's 2" (5 cm) from the edge (see page 14 for how to score). If you accidentally go all the way through, it's not the end of the world **(fig. b)**.

3. Carefully peel off the top layer and corrugation, being careful to leave the bottom layer. We'll apply glue to this section in a bit **(fig. c)**.

5. Right now we just have a flat piece of cardboard, but we need to make it into a cylinder. Using the edge of the table, start on one side and carefully press the cardboard over the edge of the table. Do this in about ½" (1.3 cm) increments until you've creased the entire sheet **(fig. d)**.

6. You should now have a nice little cylinder. Apply glue to the thin piece that you made by tearing off the top layer and make your cylinder permanent **(fig. e)**.

7. You'll notice that two of your cutouts are now covered up. Go ahead and recut those pieces **(fig. f)**.

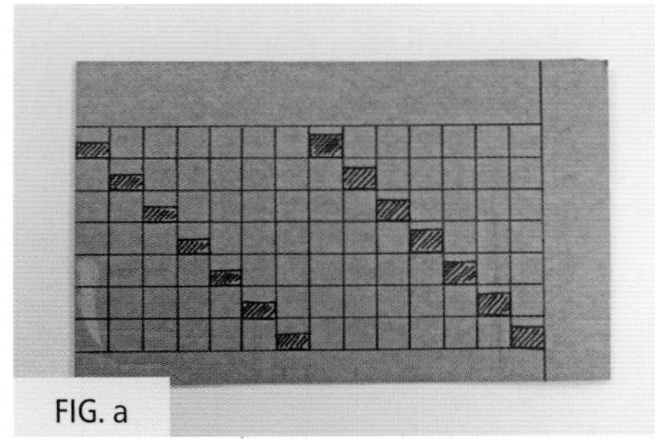
FIG. a

FIG. b

FIG. c

FIG. d

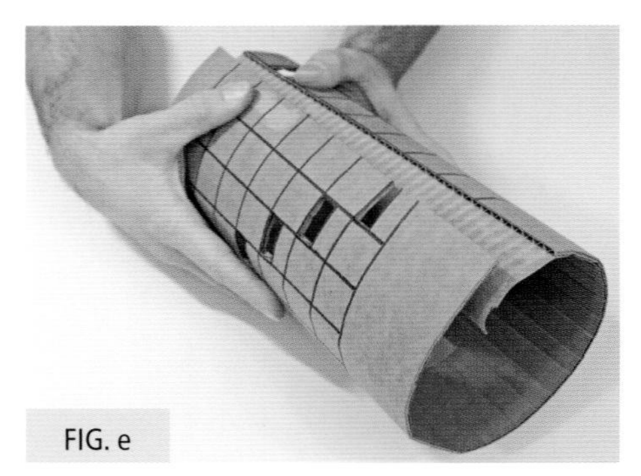
FIG. e

FIG. f

8. Place two rubber bands around every cutout, for a total of 7 pairs (14 rubber bands total). Try to space them on center for the ½" (1.3 cm) cutout and slightly toward the bottom for the ¾" (1.8 cm) cutout. A little tape secures them in place, but we can always adjust later if need be **(fig. g)**.

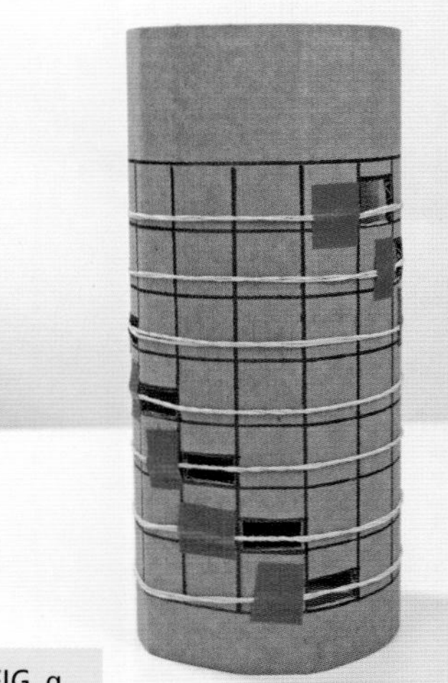
FIG. g

## Add the Knives to Make Music

1. Insert all of the knives into the cutouts. The ½" (1.3 cm) sections are for the blades of the knives, and the ¾" (1.8 cm) sections are for the handles. The goal here is to have the knives resting only on the rubber bands and not touching the cardboard. You can raise and lower the rubber bands to get it just right and then retape once you're satisfied **(fig. h)**.
2. The knives are going to act as a spiral staircase for the marbles, but we need to put up a railing so they don't fly off the sides. Gently bend C so it has some curvature and attach it to the music machine cylinder using B1, B2, B3, and B4. Be careful not to have the cardboard touching the knives. It'll dampen the sound they make **(fig. i)**.
3. Cut out a rounded section on A so it roughly matches the curvature of your music cylinder. Bend up the outer edge **(fig. j)**.
4. Glue this piece to your music cylinder so it's just above the first knife. This will be your ramp that lets the marbles flow onto the knives **(figs. k and l)**.

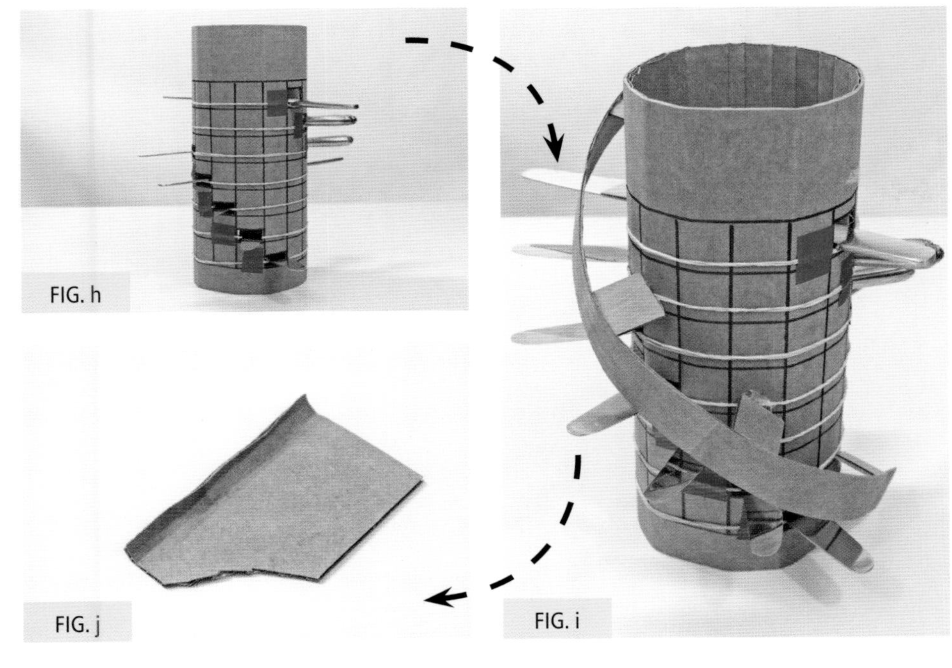

FIG. h

FIG. j

FIG. i

FIG. k

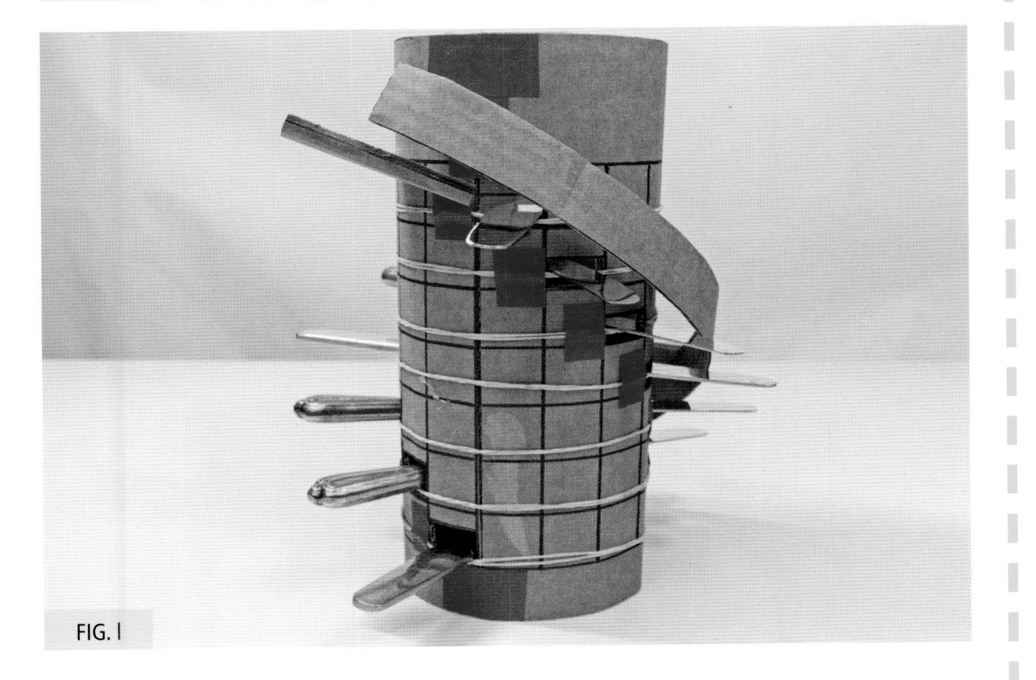

FIG. l

## Make the Marble Toy Structure

1. Glue two pairs of uprights from G1, G2, G3, and G4 **(fig. m)**.
2. Glue the uprights to the edge of the base (M) with J1 and J2 in between to act as spacers. Add the H triangular pieces for extra support **(fig. n)**.
3. Add I1 and I2 to the tops of the uprights to give them some thickness **(fig. o)**.
4. Glue together N1, N2, N3, and N4. This will be your marble lever **(fig. p)**.

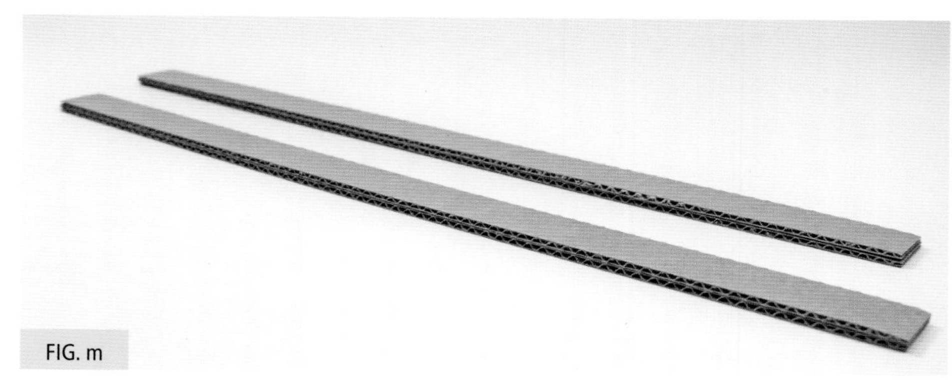

FIG. m

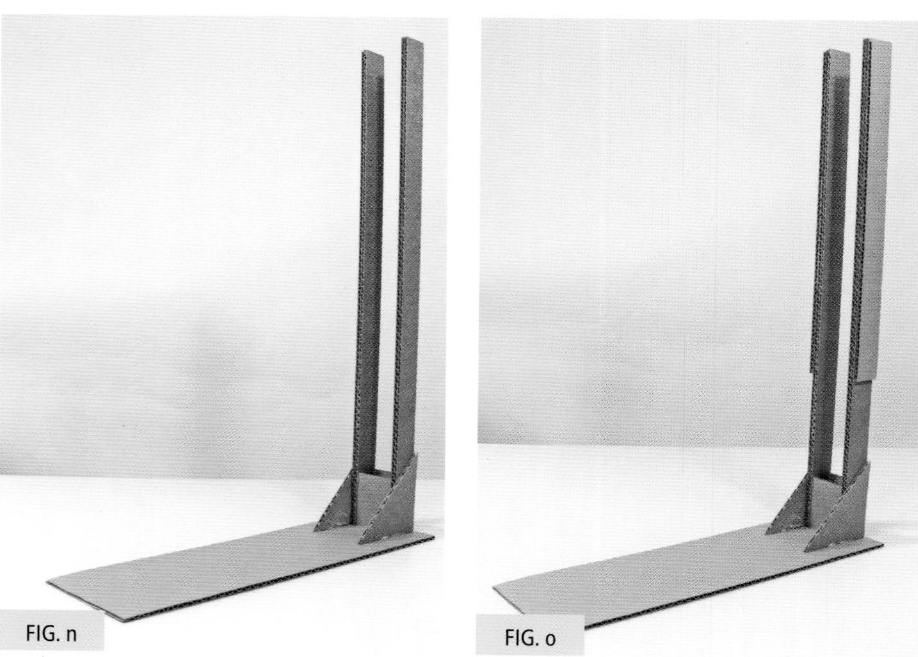

FIG. n

FIG. o

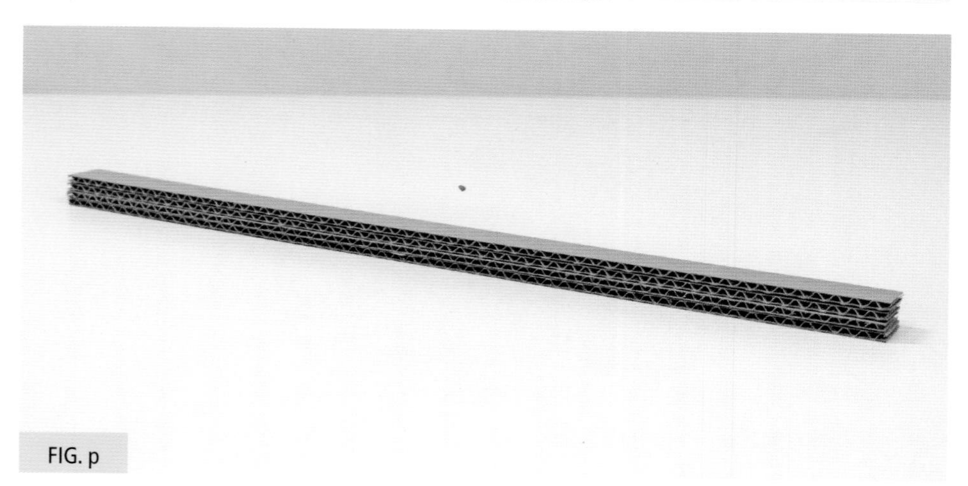

FIG. p

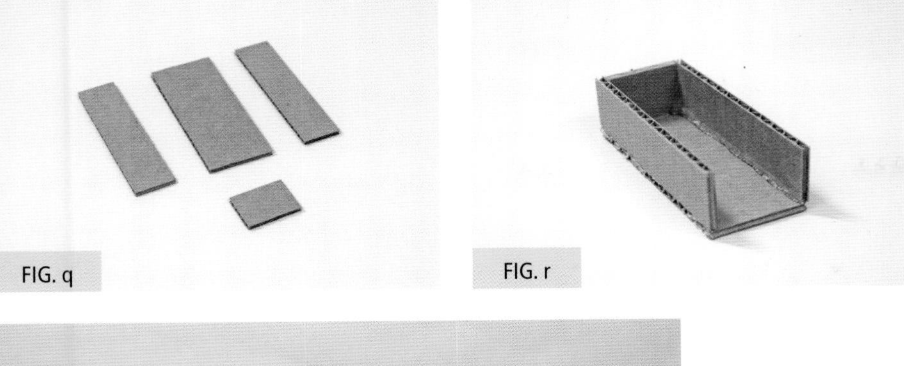

FIG. q

FIG. r

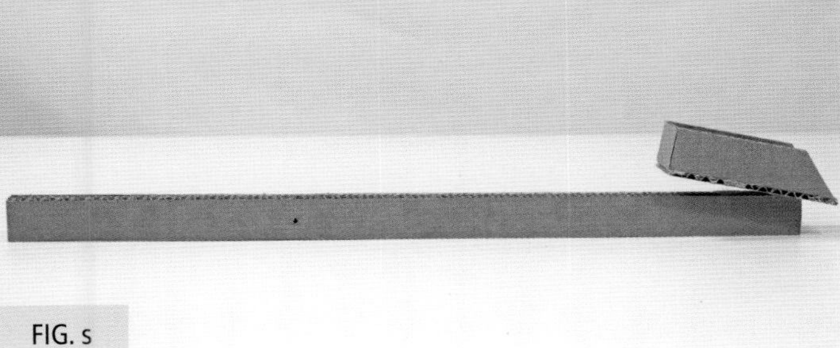

FIG. s

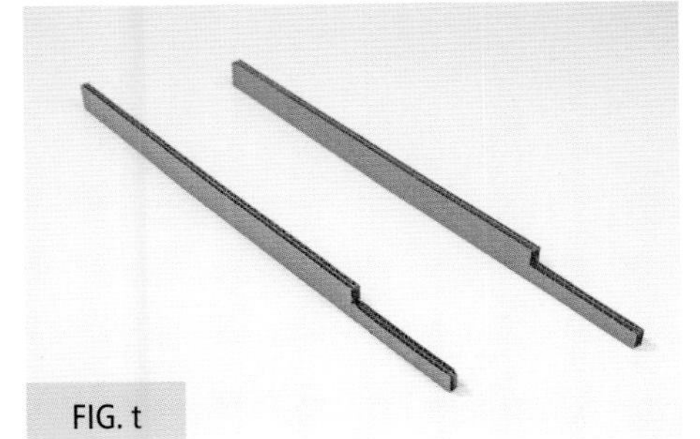

FIG. t

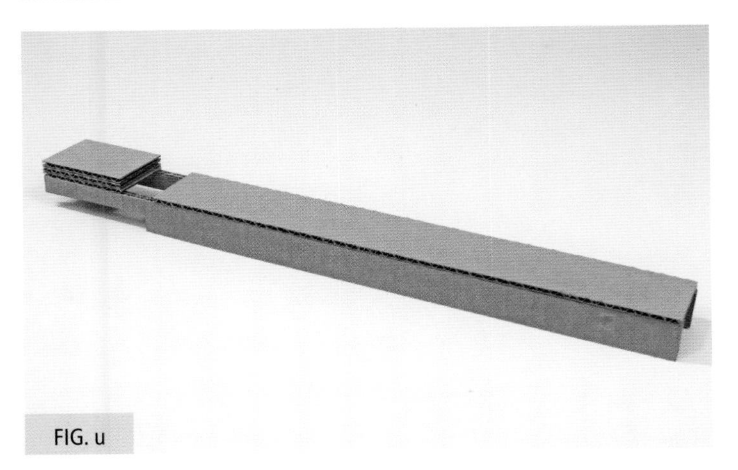

FIG. u

## Make the Trough and Rails

1. Lay out O, P1, P2, and Q as shown to make a three-sided trough **(fig. q)**.
2. This trough will attach to your marble lever and is what will lower the marbles onto the music cylinder **(fig. r)**.
3. Cut the corners off the front of your marble trough and attach it to one end of your marble lever. Don't glue it on flat. Just glue the front-bottom edge to the edge of the lever. Leave a ½" (1.3 cm) gap or so between the back edge and the lever. With an awl, poke a hole 5" (12.7 cm) from the other edge **(fig. s)**.
4. Next up we'll need to make the rails for the track where the marbles will rest. Glue S1 to S2 and S3 to S4 to make a pair of tracks **(fig. t)**.
5. Glue R to the bottom of your rails. Be mindful that the cutouts in the rails are on the opposite side. Glue together J3, J4, and J5, and attach it to the rails. Try to apply the hot glue to the long edges. We'll be poking a hole near the short edge for the smooth wire, and if there's glue in between those layers it'll get gunky and not allow the wire to move smoothly up and down. Once the glue dries, poke a hole ⅛" (3 mm) from the edge closest to the opening in the track **(fig. u)**.

### Make the Marble Release Mechanism

1. Take L1 and place it over the short piece of the rail structure. You want the spacing between the gap and the circle to be smaller than your marble, closer to half the width of your marble. Poke a hole in the center (see page 12). The poked hole should be large enough for your wire to move up and down smoothly. Repeat with L2 **(fig. v)**.
2. Glue together the two circles, making sure the holes line up. Glue in the 2" (5 cm) wire (W). Make sure this wire is perpendicular to the base of the circle. If it's crooked you might have issues later **(fig. w)**.
3. Insert the wire into the poked hole in the track and glue two of the marbles to the bottom. If the glue doesn't feel like it's going to hold, you can add some tape. Make sure you don't accidentally glue the marbles to the rail structure. It needs to move up and down freely, but don't test until your hot glue has cooled down **(fig. x)**.
4. Glue the rail structure to the top of the uprights so it's at a slight angle. Make sure the orientation is the same as the picture **(figs. y and z)**.
5. Poke a hole into the uprights 7" (17.8 cm) from the top of the rail. Use your pokey pad (see page 11) so you don't bend the uprights. This hole should be large enough so your toothpick can rotate freely.
6. Attach your marble lever to the towers with a toothpick (Y). Add a dab of glue to the lever and toothpick but not the towers. Cover the ends with K1 and K2 so the toothpick can still rotate freely but not slide out. Lift up the trough until it just touches the bottom marble. Use a scrap piece of cardboard as a shim to make the trough level, and glue it in place **(fig. aa)**.

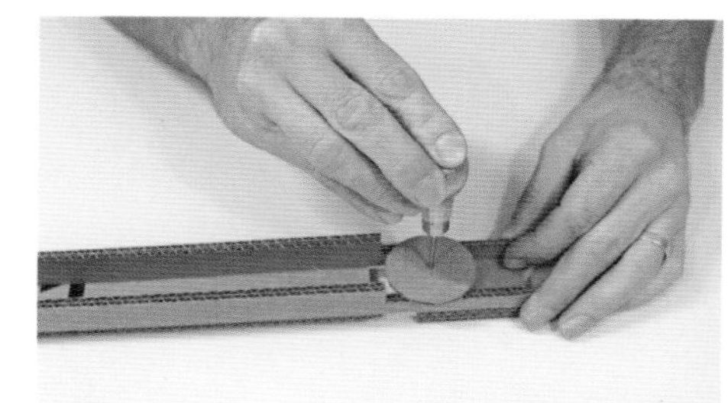

FIG. v

FIG. w

FIG. x

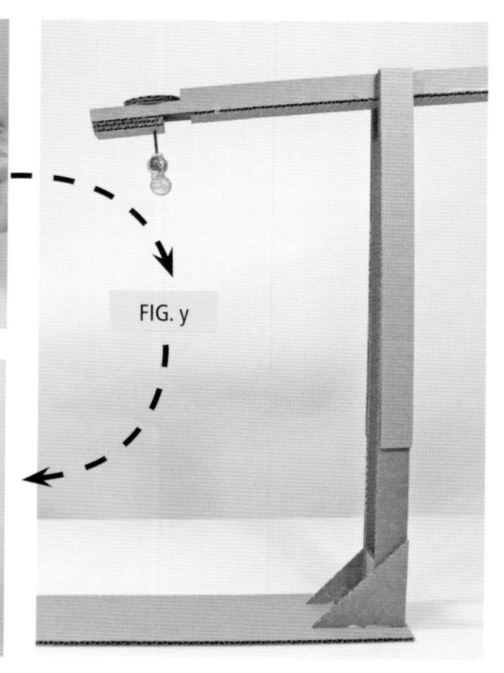

FIG. y

FIG. z

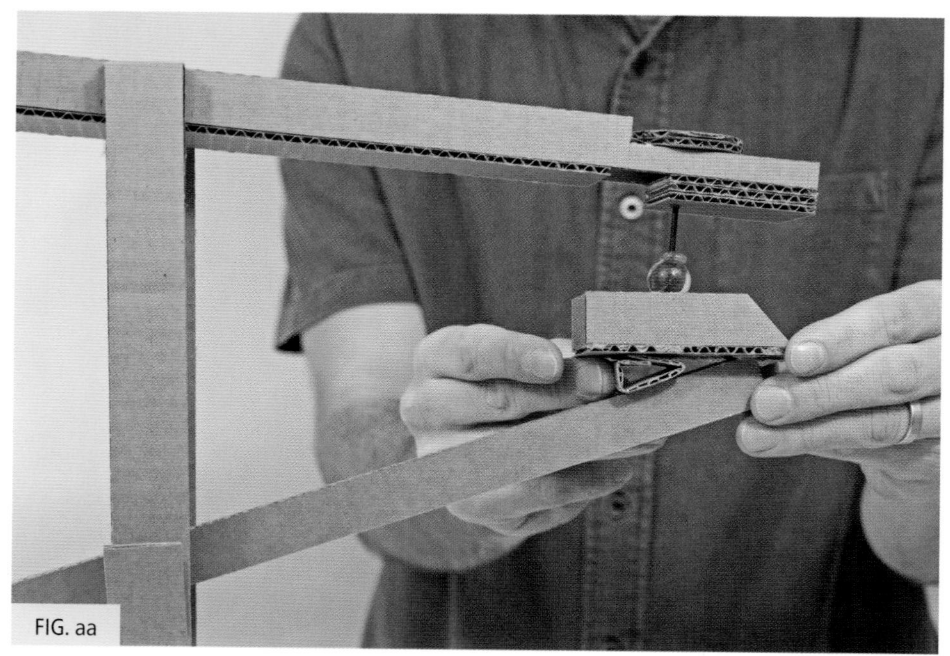

FIG. aa

## Make the Marble Counterweight

1. Lay out T1, T2, U1, U2, V1, and V2 to make the marble counterweight. You can attach tape to them while they are flat and then fold it all together **(fig. bb)**.

2. Once you fold the counterweight together, secure it with more tape **(fig. cc)**.

3. A counterweight is nothing without a little weight, so add some marbles (X) to give it some mass. I used six marbles, but you might have better luck with more or fewer **(fig. dd)**.

4. The counterweight should slide onto the end of the marble lever. If should be a tight fit **(fig. ee)**.

5. If it feels loose, you can glue a cardboard shim to the lever **(fig. ff)**.

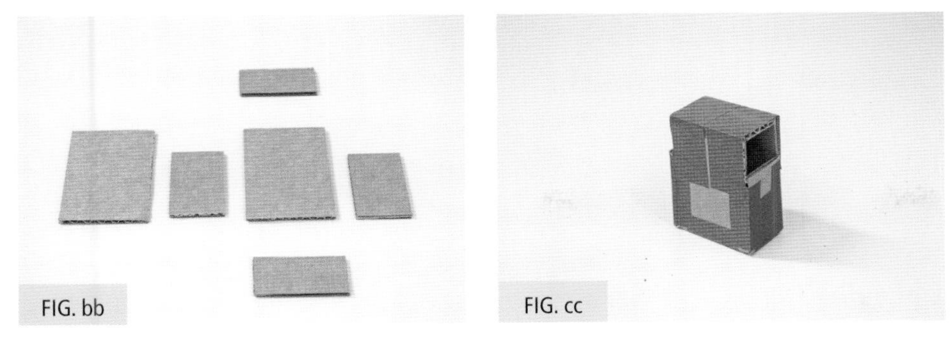

FIG. bb

FIG. cc

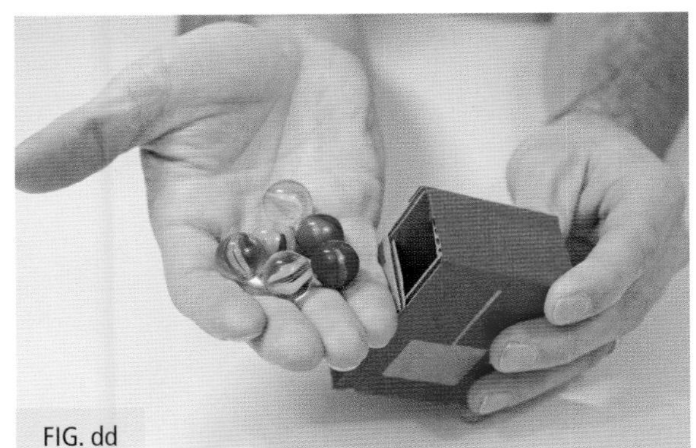

FIG. dd

FIG. ee

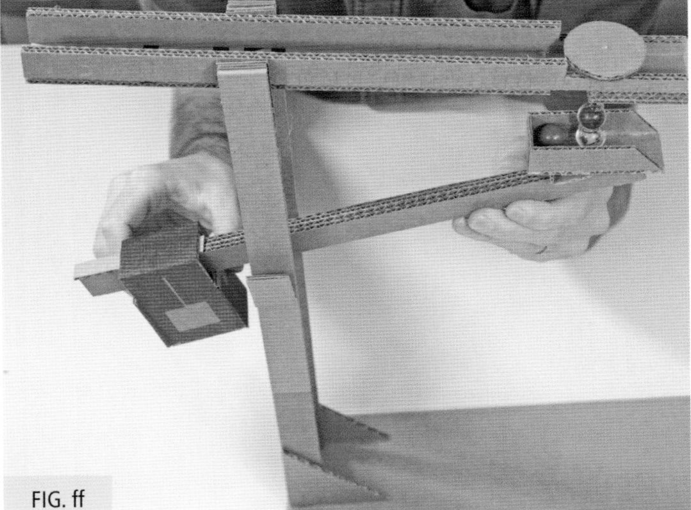

FIG. ff

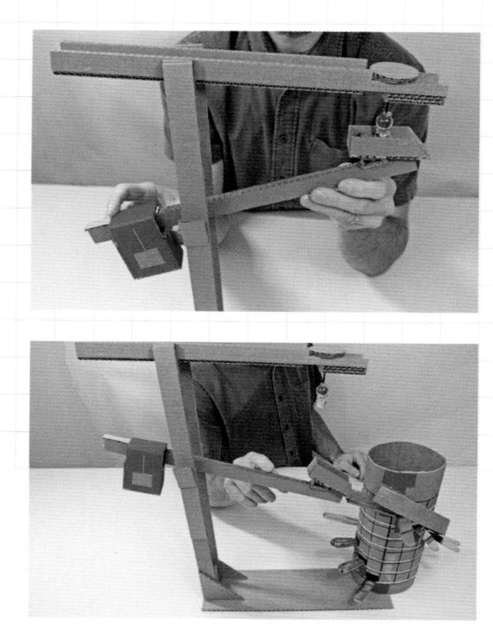

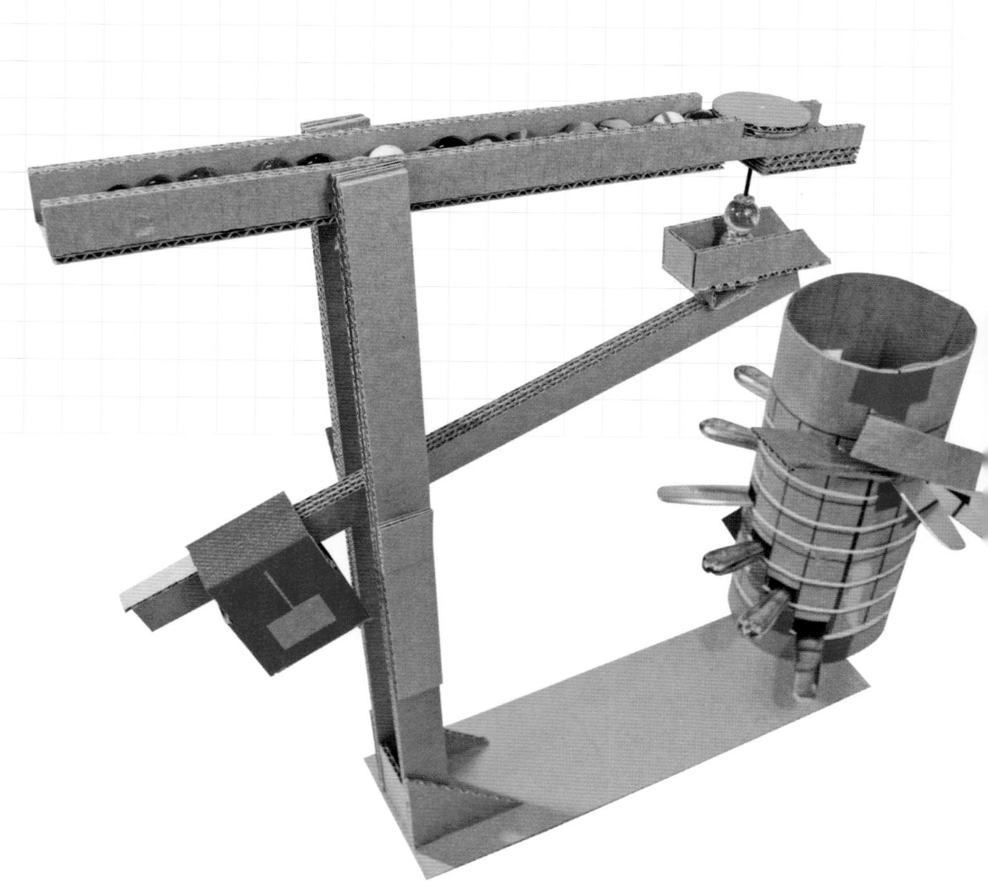

# START THE CHAIN REACTION!

### The Setup

1. This counterweight is just like the old scales at the doctor's office. Place a marble into the trough and move the counterweight toward the center until the trough lowers.
2. Bring in your music cylinder and line up the ramp with the trough. The goal is to have a marble fall onto the ramp.

### Make It Go!

Fill up your marble track with marbles that are the same size and weight. That's important. Lift up on the lever on the counterweight end and gently release. The trough should bump the two marbles attached to the wire and circular pieces. This lifting motion should allow one marble to pass through. That marble falls into the trough, which lowers onto the ramp. The marble goes down the ramp, creating musical sounds, while the trough rises back up and bumps the mechanism again, restarting the process. The machine will run until you're out of marbles.

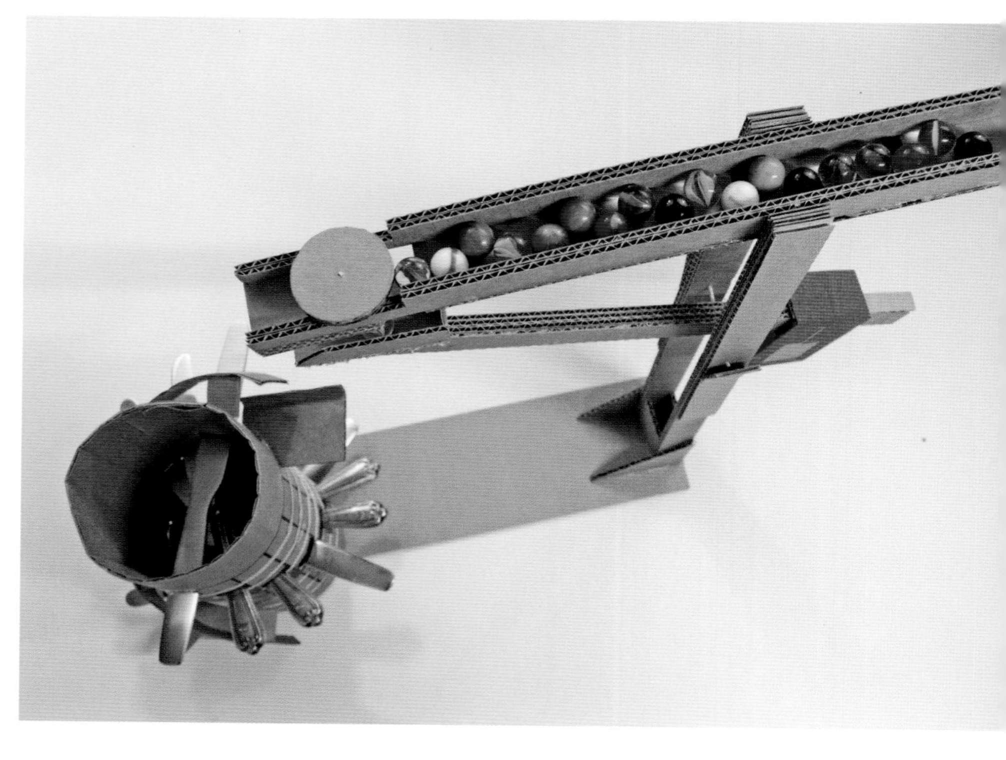

### Troubleshooting

If too many marbles are falling through the cutout at once, or they're getting stuck, you may need to adjust the length between circular pieces and the edge of the gap. If the gap is too large, you can add material around the circular pieces to make them bigger, or better yet, just cut a new one. If the gap is too small, you can carefully cut a little bit of cardboard until it's just right **(fig. a)**.

You also might be having problems with the wire getting jammed and not going up and down smoothly. If this is the case, make sure there is no glue or anything tacky on the wire. If you got too much hot glue in that area when you glued together the three pieces, you might want to pop it off and make another cardboard sandwich to replace it.

If it's still not working, you might have to add or take away marbles in your counterweight, or move it around until it's just right. Even if the fit is snug, it might be moving around, so once you find the right spot you can lock it into position with a piece of tape **(fig. b)**.

FIG. a

FIG. b

### Engineering Know-How

The marble machine works because of the counterweight. It's just heavy enough to keep the lever in the up position. When the marble hits it, it shifts the center of gravity to the other side of the pivot point, which causes the lever to drop down, letting the marble roll into the music machine. The moment the marble rolls off the lever, the center of gravity shifts back to the other side of the pivot point, causing the lever to go back to its original position.

# BALLOON POPPER

If you're like me, you flinch anytime you know a balloon is about to pop. Why not increase that anxiety by building a machine that pops a balloon?

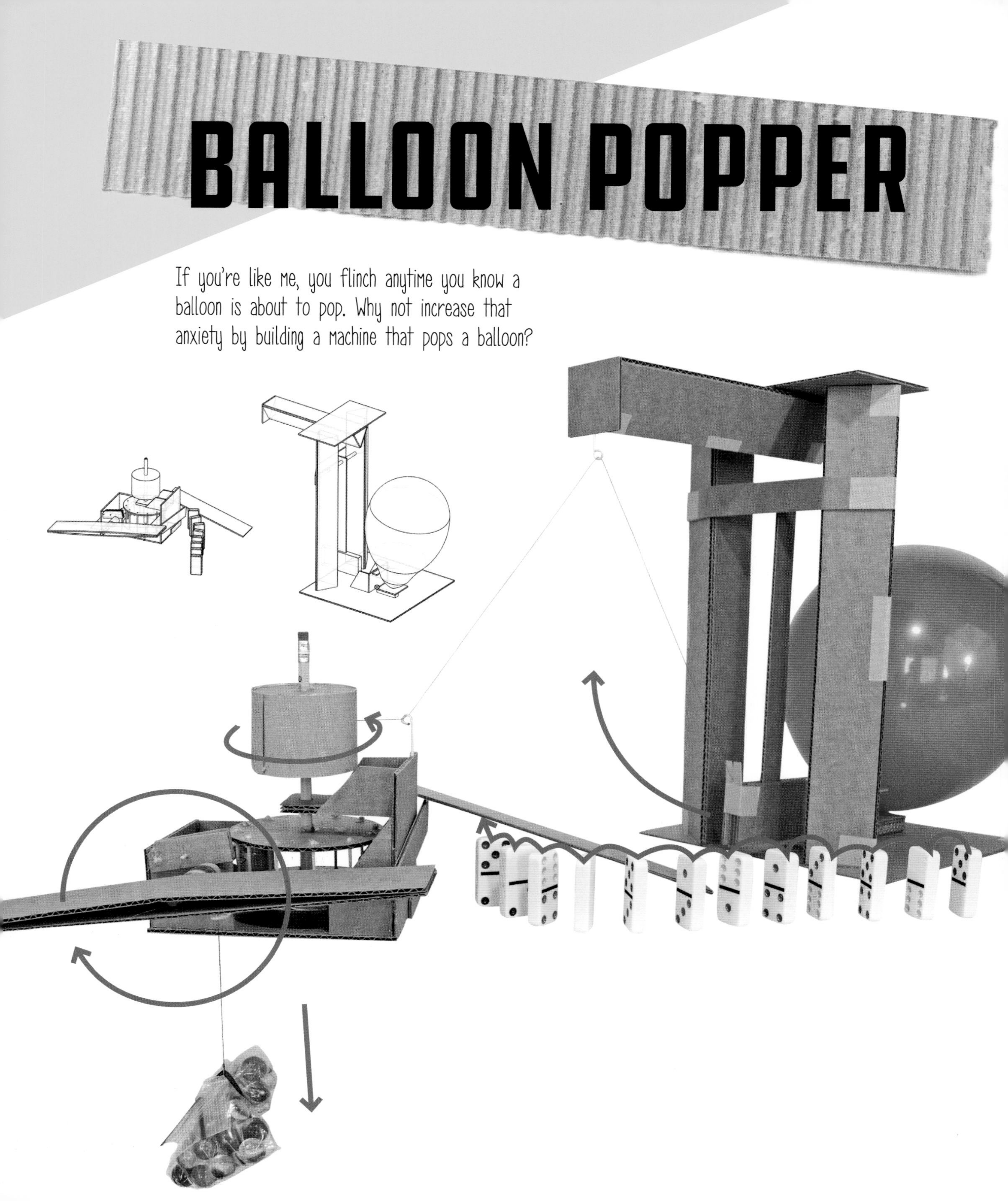

# BUILD MATERIALS & TOOLS

### MAIN STRUCTURE

**CARDBOARD:**

Two 6" x 16" (15.2 x 38 cm), scored lengthwise at 2" and 4" (5 and 10.2 cm) (see page 14)–**A1 & A2**

One 1" x 8" (2.5 x 20.3 cm)–**B**

One 4" x 8" (10.2 x 20.3 cm)–**C**

One 10" x 10" (25.4 x 25.4 cm)–**D**

One 6" x 10" (15.2 x 25.4 cm) scored lengthwise at 2" and 4" (5 and 10.2 cm) (see page 14)–**E**

One 2" x 2" (5 x 5 cm)–**F**

**OTHER:**

1½" (3.8 cm) piece of wire, bent into loop on one end–**G**

### BALLOON HOLDER

**CARDBOARD:**

One 2" x 2" (5 x 5 cm), scored every ½" (1.3 cm)–**H**

One 2" x 2" (5 x 5 cm)–**I**

One ½" x 3" (1.2 x 7.6 cm)–**J**

One ¾" x 1½" (1.8 x 3.8 cm) piece of cardstock or thin cardboard with notch and slit cut in top–**K**

Two 2" x 2" (5 x 5 cm) with 1" x 2" (2.5 x 5 cm) triangle removed–**L1 & L2**

Two 1½" x 2" (3.8 x 5 cm)–**M1 & M2**

One 1" x 12" (2.5 x 30.5 cm)–**N**

One 2" x 3" (5 x 7.6 cm) with hole poked in center ½" (1.3 cm) from the top–**O**

Four 4" x 2" (10.2 x 5 cm)–**P1, P2, P3 & P4**

**OTHER:**

Balloon–**Q**

Needle–**R**

Toothpick, cut to 1" (2.5 cm)–**S**

Piece of paper, ¾" x 5" (1.8 x 12.7 cm)–**T**

*(continued)*

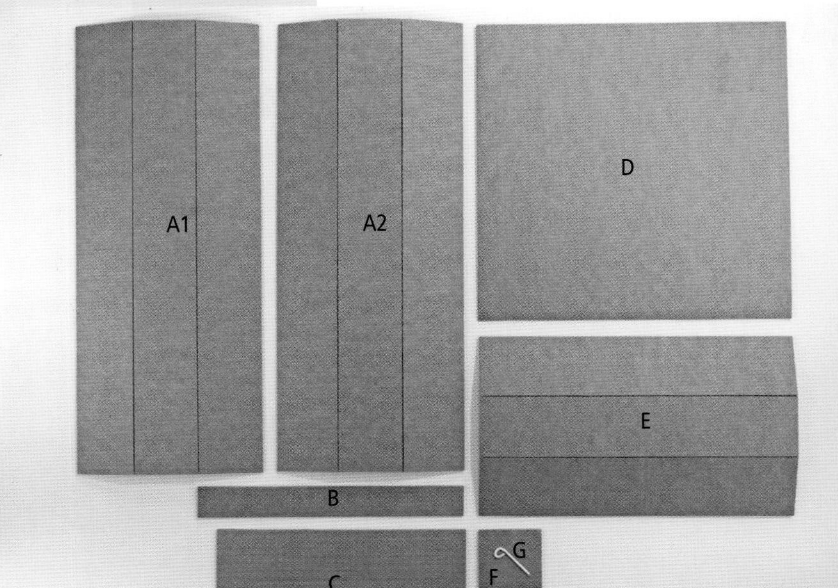


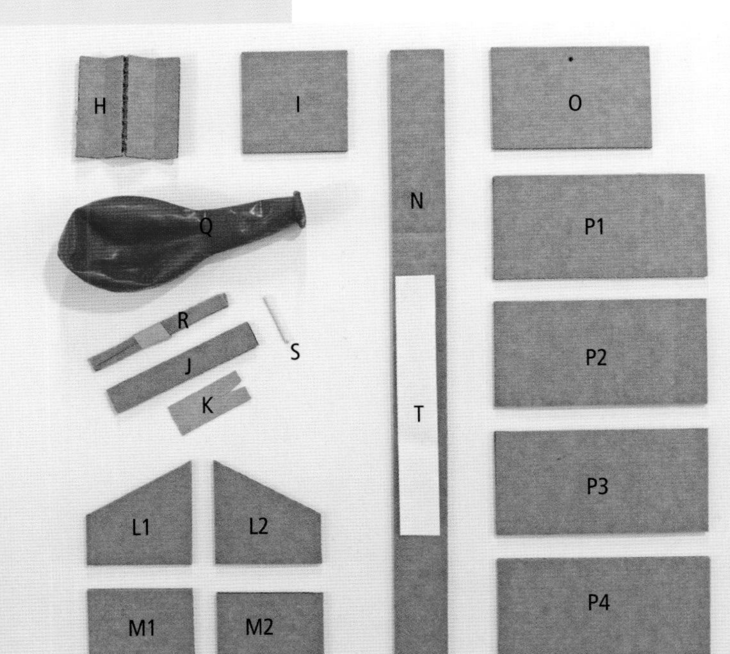


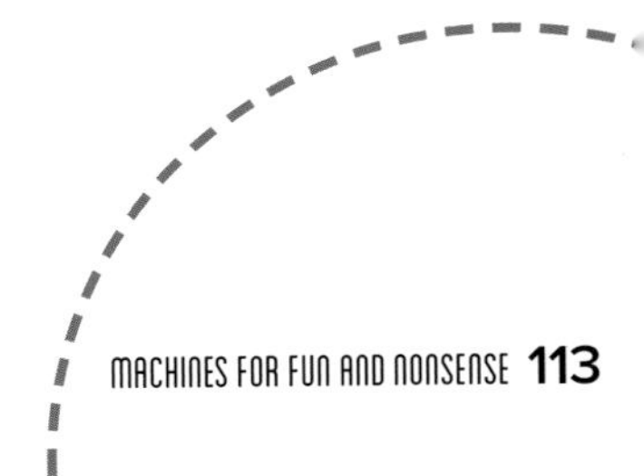

## REGULATOR AND LANTERN GEAR

### CARDBOARD:

Two circles, 2½" (6.4 cm) in diameter (see page 15)–**U1 & U2**

One 2" x 4" (5 x 10.2 cm)–**V**

One 1" x 5" (2.5 x 12.7 cm)–**W**

One 1" x 6" (2.5 x 15.2 cm) with corner cut out–**X**

One ½" x 2" (1.3 x 5 cm)–**Y**

One ½" x 3" (1.3 x 7.6 cm)–**Z**

Two ¾" x 2" (1.8 x 5 cm)–**AA1 & AA2**

Two 2" x 2" (5 x 5 cm) with hole poked in corner ½" (1.3 cm) from edges–**BB1 & BB2**

Two 1½" x 3¼" (3.8 x 8.3 cm) with a long triangle cut off the end–**CC1 & CC2**

One 2" x 2" (5 x 5 cm) with a notch cut out ½" (1.3 cm) from the edge–**DD**

Two 1½" x 3" (3.8 x 7.6 cm) with hole poked in corner ½" (1.3 cm) from edges–**EE1 & EE2**

One 4" x 4" (10.2 x 10.2 cm)–**FF**

Two 1½" x 4" (3.8 x 10 cm) with a long triangle cut off the end–**GG1 & GG2**

One 4" x 7" (10.2 x 17.8 cm)–**HH**

Two circles, 4" (10.2 cm) in diameter (see page 15) with with 12 holes poked evenly ½" (1.3 cm) from edge–**II1 & II2**

Two 2" x 11" (5 x 28 cm)–**JJ1 & JJ2**

2" x 9" (5.1 x 22.9 cm)–**KK**

### OTHER:

Plastic bag (to hold weights, like marbles)–**LL**

20 to 25 marbles to use as a weight–**MM**

Thin plastic, like a yogurt lid–**NN**

Two 1½" (3.8 cm) wire with loop on one end–**OO1 & OO2**

Small rubber band–**PP**

6' (2 m) strong string–**QQ**

2 pencils–**RR**

14 toothpicks–**SS**

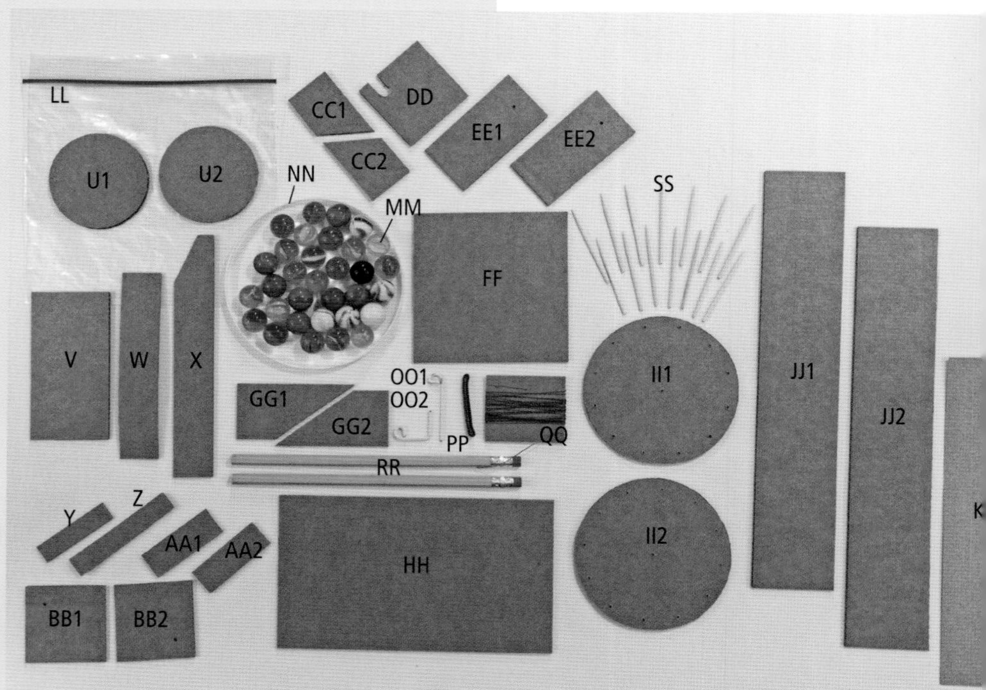


## NOT SHOWN

Dominoes

## TOOLS

Awl

Ruler

Sharpened pencil

Painter's tape

Scissors

Utility knife

# BUILD THE MACHINE

### Make the Tower Structure and Balloon-Popping Hammer

1. With an awl, poke a hole 3" (7.6 cm) from the top edge on A1 and A2. If you made a pokey pad (see page 11), go ahead and use it! Using a sharpened pencil, carefully expand the holes **(fig. a)**.
2. Fold the towers into their triangular shapes and secure with tape **(fig. b)**.
3. To make the hammer, start with N and poke a hole ½" (1.3 cm) from the top edge. Use a pencil to enlarge the hole. Be sure to make this hole slightly larger than the pencil so it can spin freely **(fig. c)**.
4. Glue together P1 and P2. Add the handle (N) and glue M1 and M2 on either side. Complete the sandwich by adding P3 and P4 on the outside **(fig. d)**.
5. Insert the pencil through the hole in the handle. Place a tower on each end, making sure that the pencil end fits into the corners of the triangle **(fig. e)**.

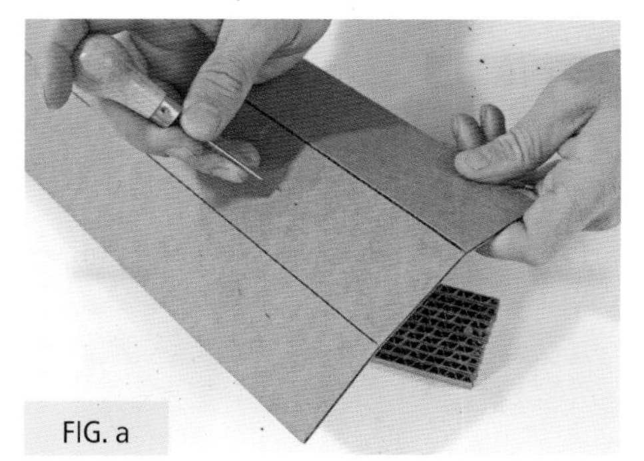
FIG. a

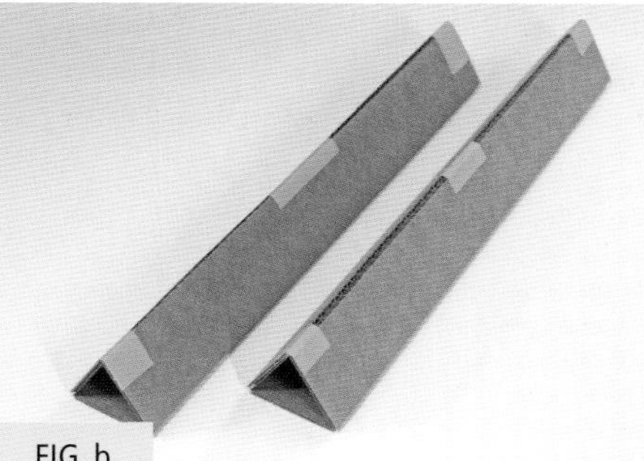
FIG. b

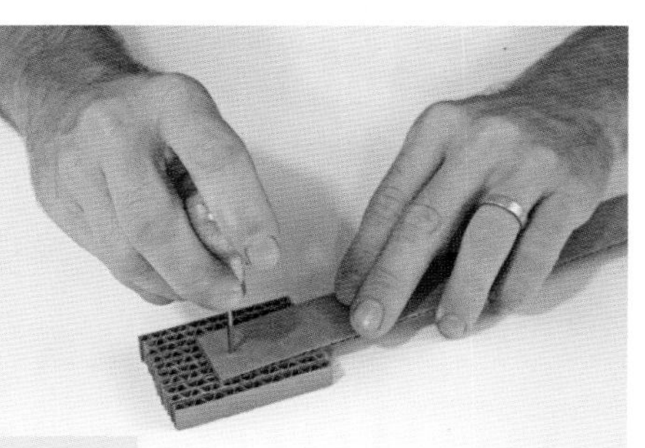
FIG. c

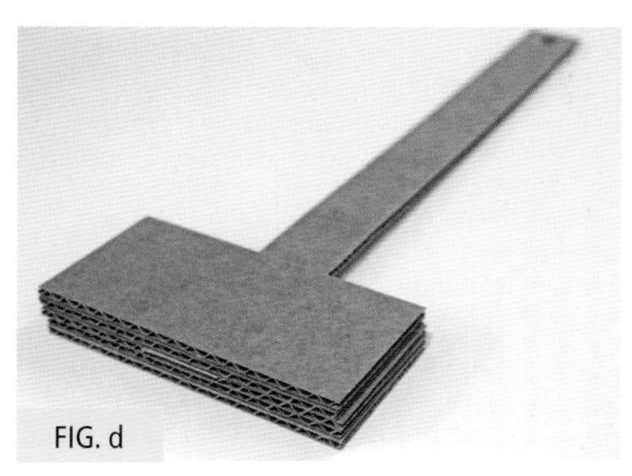
FIG. d

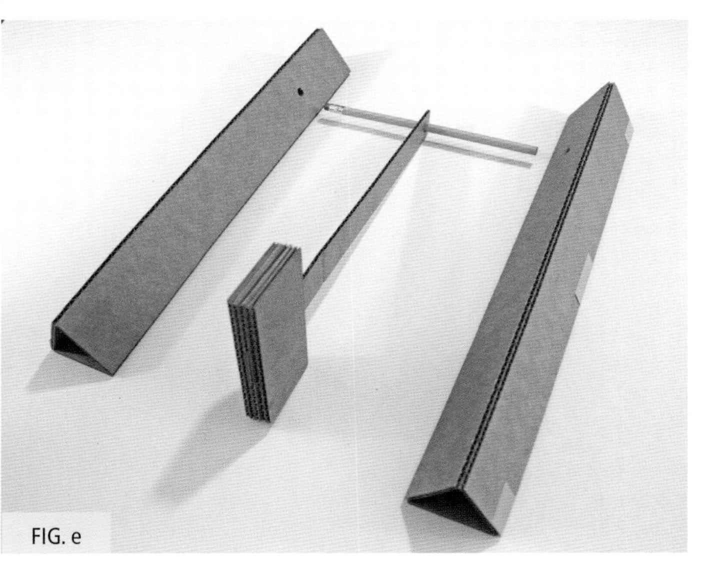
FIG. e

6. Add tape on either side of the handle to keep it centered between the towers **(fig. f)**.
7. Glue the towers to the edge of the base (D). The spacing between the towers isn't super important, but if you're using a fresh pencil it'll be about 4½" (11.4 cm). Just try to keep your towers straight and perpendicular to your base **(fig. g)**.
8. Assemble E into a triangular tube **(fig. h)**.
9. Attach it to the center of C. Glue F to the end **(fig. i)**.
10. Glue the triangular tube structure onto the tops of the towers, making sure it's facing *away* from the base. Add a looped piece of wire (G) to F. This will act as a guide for the string **(fig. j)**.
11. Tape B about 3½" (9 cm) from the top of the tower. Don't use glue because this piece will need to be adjusted **(fig. k)**.
12. Attach K to the end of the hammer on the side away from the base. This will hold the string that will raise the hammer **(fig. l)**.

### Make the Cage (or Lantern) Gear

To make the cage gear, divide II1 into 12 sections (refer to page 16 on how to divide circles). Poke holes ¼" (6 mm) from the edge of the circle at the division marks. Use this circle as a template to poke holes in II2. Enlarge the center hole with a pencil, and insert toothpicks, 12 in all, in the poked holes of one circle. Carefully and methodically insert the free ends of the toothpicks into the corresponding holes of the other circle. Slide the pencil through the enlarged holes, and shift and twist the circles until all of the toothpicks are evenly pushed into the holes and are perpendicular to the circles. The distance between the circles should be about 2" (5 cm). Once you're satisfied, trim the ends and add glue to the toothpick tips and between the pencil and cardboard **(fig. m)**.

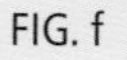
FIG. f

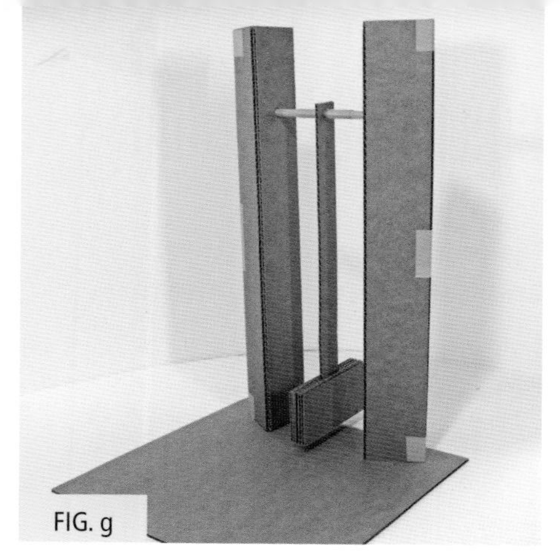
FIG. g

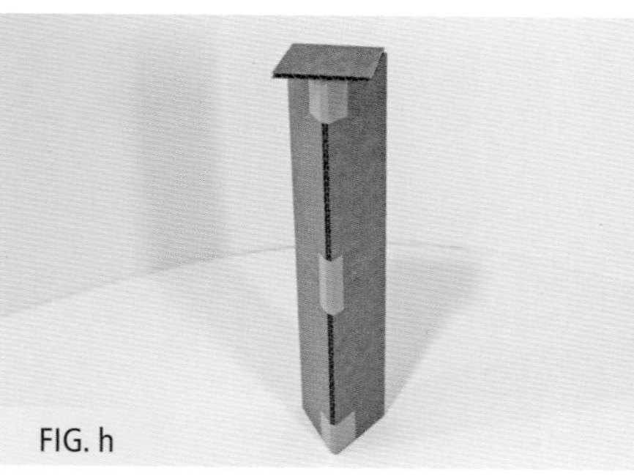
FIG. h

FIG. i

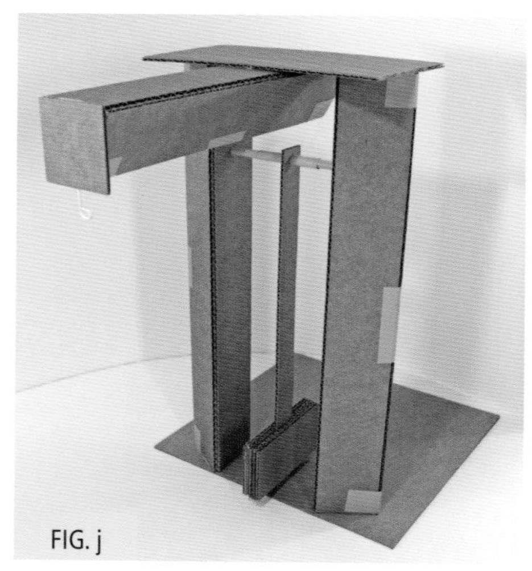
FIG. j

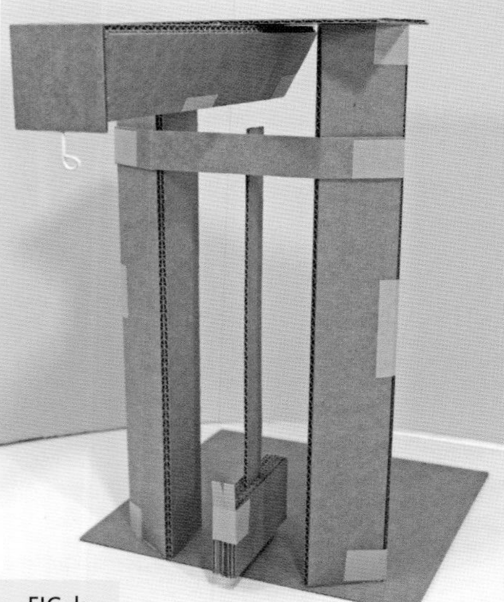
FIG. k

FIG. l

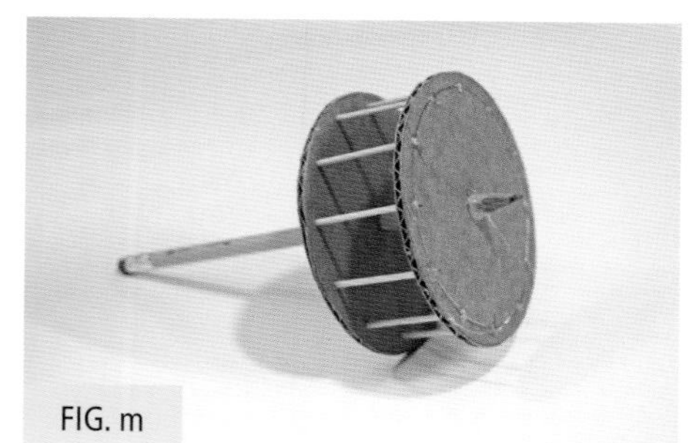
FIG. m

## Make the Cage Gear Housing

1. Glue FF perpendicular to the edge of HH. Add GG1 and GG2 as side supports **(fig. n)**.

2. Glue together the EE1 and EE2 and poke a hole in the combined pieces ½" (1.3 cm) from the edge. Enlarge the hole with a sharp pencil and slide this over the pencil in your gear, making sure the pencil can rotate freely. Keeping the pencil as vertical as possible, make a mark on the base. Gently push the pencil to poke a hole, or use an awl or sharp needle. Your gear will rotate about this point. Secure in place with glue. Now glue EE to the L-shaped structure. Add CC1 and CC2 as supports above it to make it more secure **(fig. o)**.

3. Find the center of U1 and U2 (see page 12). Poke a hole in the center and enlarge with a sharp pencil. Make a 2" (5 cm) tall cylinder with the two circles and KK (see page 17). Trim the strip and glue in place. Slide it over the pencil and glue it in place. Make sure it's not touching the supports you just added **(fig. p)**.

4. Make a compass (see page 15) and draw a 1" (2.5 cm) circle on your plastic lid. Draw a ½" (1.3 cm) circle inside that. Cut out this donut shape by cutting across to the center of the circle and then cutting around the drawn lines. This piece will act as your worm and will turn the lantern gear **(fig. q)**.

5. Using another sharpened pencil, carefully tape the plastic worm to the shaft of the pencil, about 1½" (3.8 cm) from the top. If you hold the pencil with the sharpened side up, the worm will go to the left if the top is the starting point. Try to keep the plastic tight against the pencil and make sure the ends are secured well. Depending on the plastic you use, glue may or may not stick, but once you're happy with the location, you can add glue to the ends to prevent the tape from coming undone **(fig. r)**.

FIG. n

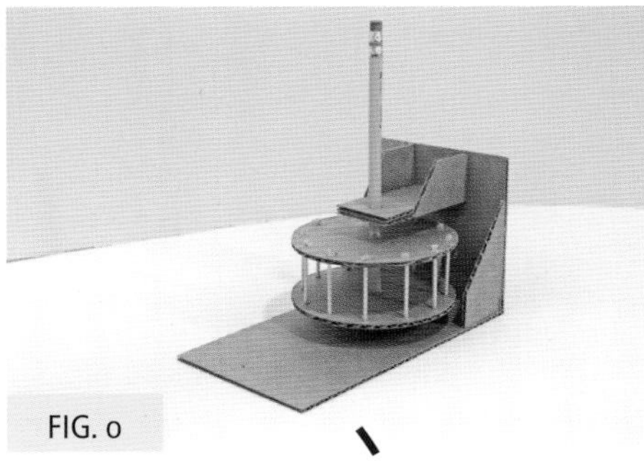
FIG. o

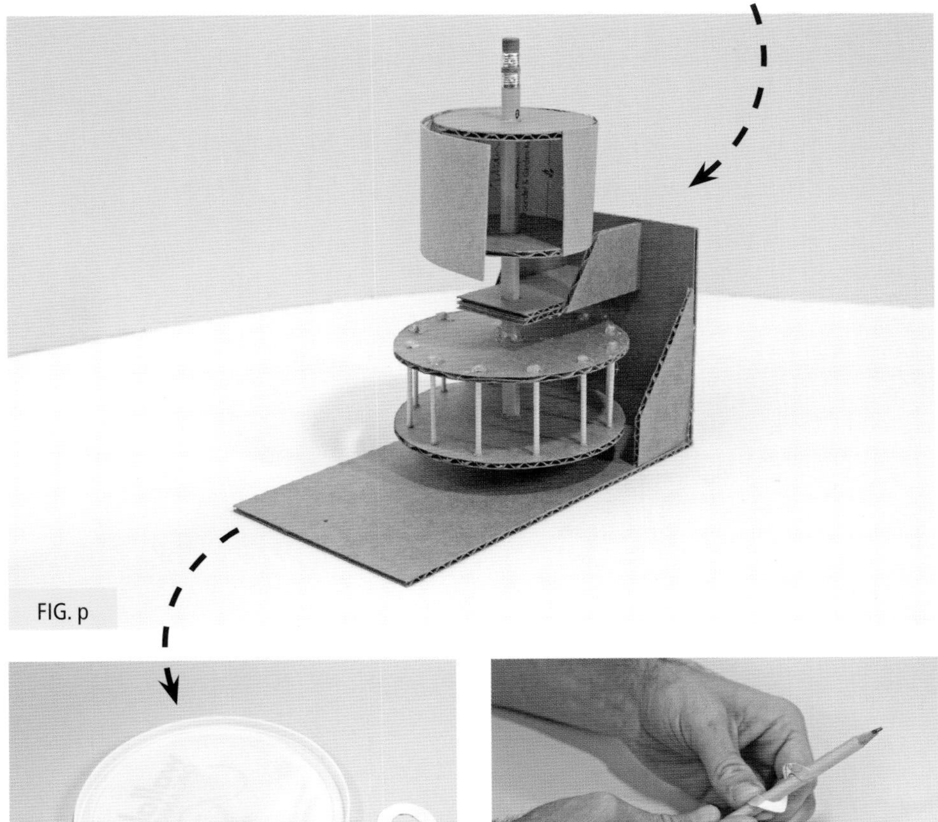
FIG. p

FIG. q

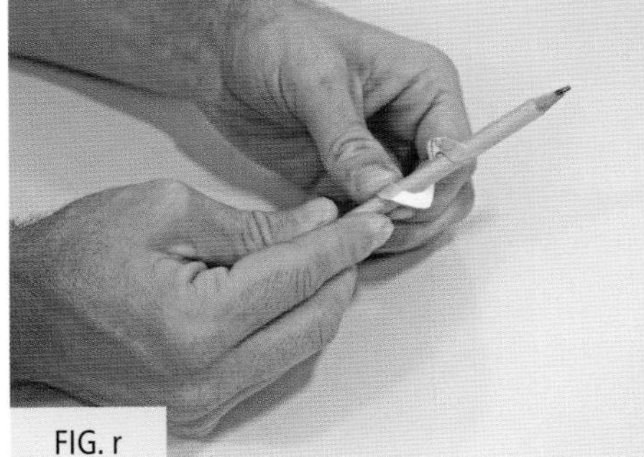
FIG. r

6. Glue together BB1 and BB2. Stick the tip of the pencil with the worm on it in the hole. Using your eye, move it so the pencil is tangent to your lantern gear and parallel with the front edge of the base. Twist the pencil clockwise a few times to get a feel for how smooth it turns the gear and if there's too much wiggle room. Once you find the sweet spot, glue the BB pieces in place **(fig. s)**.
7. Assemble the shaft holder by sticking a toothpick through the end of AA1 and AA2, creating a hinge. This will hold the shaft of your worm in place but allow you to easily remove it to make any adjustments **(fig. t)**.
8. Once it's in the right spot, glue the bottom of DD, as shown in **fig. z**. Lower the hinge lever, locking the shaft in place. Poke a hole through all three pieces so you can add a pin, which will lock everything in place. To add support, attach V to DD and BB **(figs. u and v)**.
9. The worm shaft won't stay in place unless there are spacers to keep it from wiggling around. Wrap two pieces of thin cardboard around the shaft: Z on the inside and W on the outside. W will also be used to wrap a string around to turn the shaft. Secure with tape **(fig. w)**.

### Make the Regulator

1. Right now, if we attached a falling weight to the worm shaft, it would cause it to spin really fast. Sometimes that's a good thing, but for this machine we want to see all of the pieces in action. In order to slow it down, we're going to add a paddle to the shaft. This paddle is called a *regulator* because it regulates how fast the shaft will spin.
2. Sandwich JJ1 and JJ2 around the end of the shaft, making sure they're centered. Cut a length of string that's 36" (1 m) long and attach to W. Tie a rubber band to the end of it. The rubber band will be used to attach your weight. I'll be using a bag of marbles **(fig. x)**.

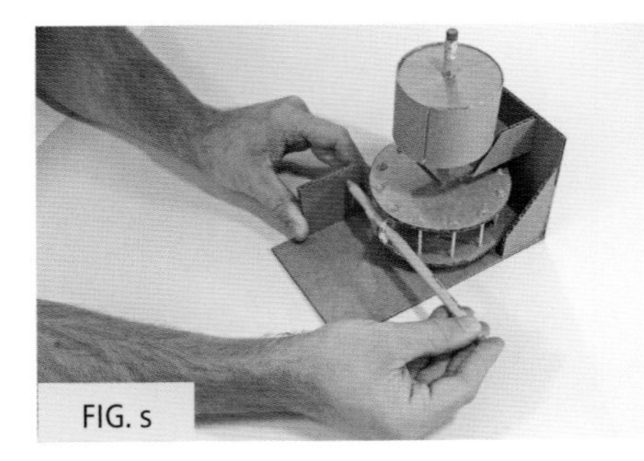
FIG. s

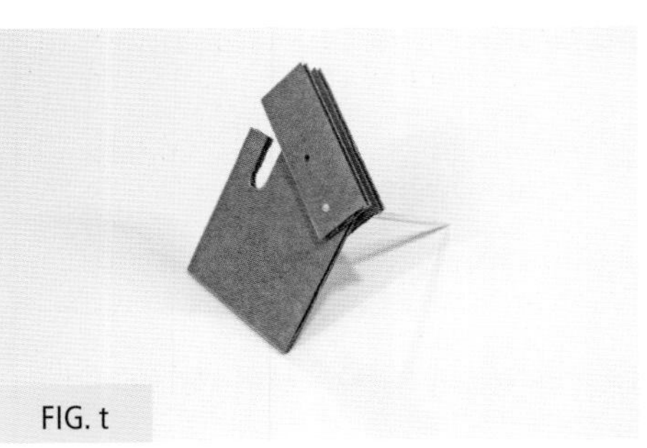
FIG. t

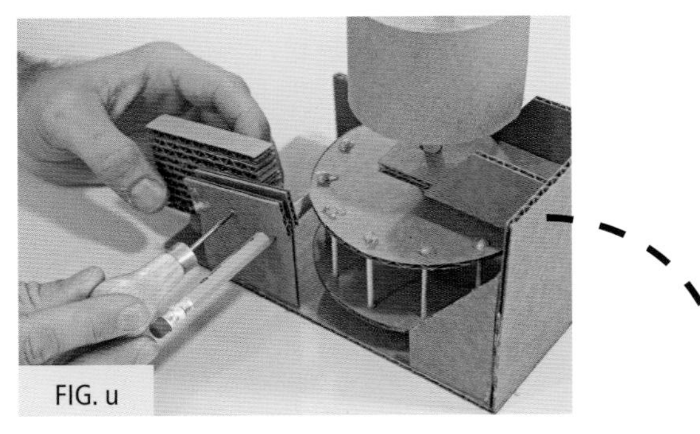
FIG. u

FIG. v

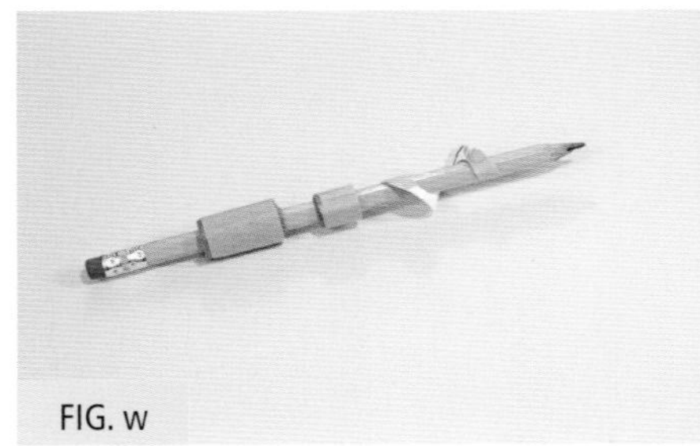
FIG. w

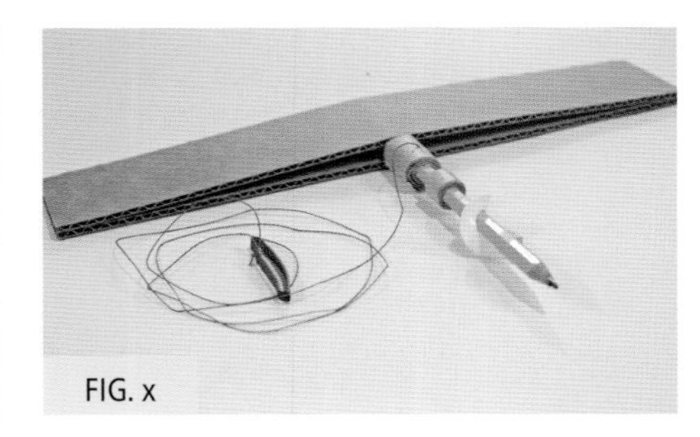
FIG. x

3. Add a looped piece of wire on the edge to help guide your string **(fig. y)**.
4. Almost there! Next we'll add X as a lever that will keep the paddle from spinning until we want it to spin. Add the lever on the other side of the structure so just the minimum amount is underneath the paddle, with the cutoff side facing down. We want the right side of the lever to be longer so we can add a light weight, like a piece of cardboard, to it so it won't let the paddle move until the piece of cardboard gets knocked off. To keep the lever from falling too far on the right side, add Y as a stopper underneath the lever.
5. Now's a good time to tape the remaining 36" (2 m) string to the 2" (5 cm) cylinder. Add another wire loop so you can help guide the string **(fig. z)**.

### Attach the Needle Popper

1. We could just stick a sharp needle on the end of our hammer and be done with it, but there's no elegance in that. Instead, we'll make a system that hides the needle until it's needed. This does two things for us: One, we don't accidentally poke ourselves. And two, we don't accidentally pop the balloon.
2. Start by gluing L1 and L2 to O. Attach the ½" x 3" (1.3 x 7.6 cm) piece to the front, trying to line up the holes in the corrugated cardboard with the hole you poked **(figs. aa and bb)**.
3. Cut the last remaining toothpick to 1½" (3.8 cm). Tape a sharp needle to the end. Glue the end of the cut toothpick to the center of the ¾" x 5" (1.8 x 12.7 cm) piece of paper (T). Fold the ends so you have a surface to apply glue to, line up the needle with the hole, and then attach the ends with glue. This creates a spring that keeps the sharp point of the needle hidden until it is pressed against **(fig. cc)**.

FIG. y

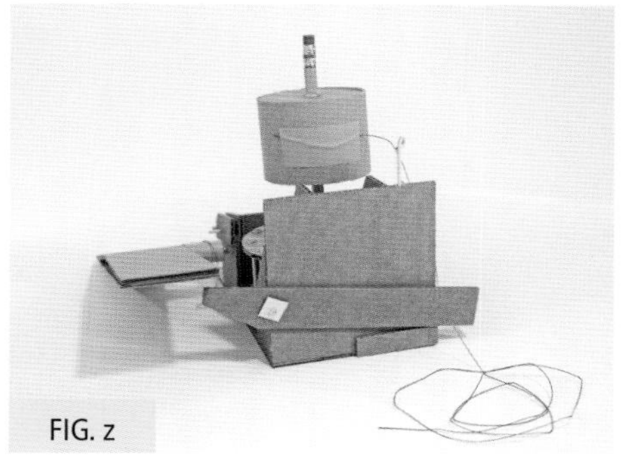
FIG. z

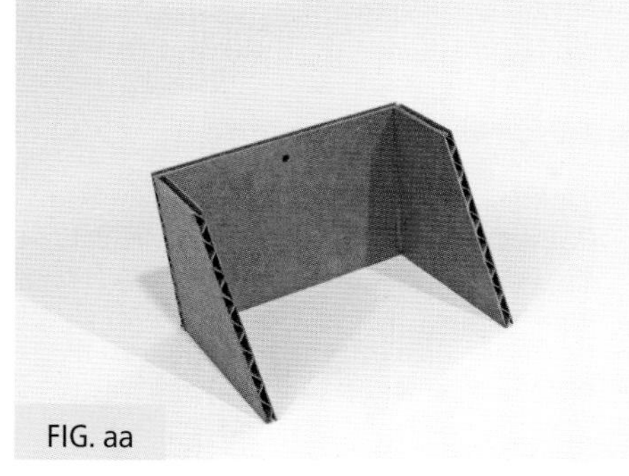
FIG. aa

FIG. bb

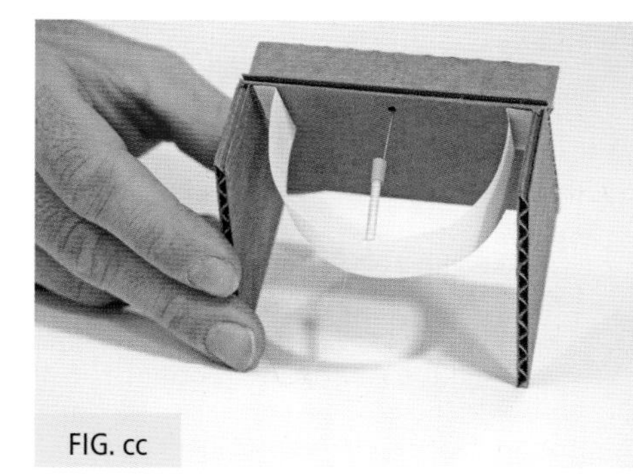
FIG. cc

4. Glue the needle structure to the base with the hammer. Attach it so it's about ½" (1.3 cm) away from the hammer when it's in a resting position. Try to center the toothpick with the middle of the hammer as best you can **(fig. dd)**.

5. Now we need a way to hold our balloon in place so we can pop it. To do that, use a utility knife to cut a slot in I. Fold and glue H to create a ½" x 2" piece that is four layers thick. Glue this to the edge of I, on the opposite side of the slit. Attach this to the base of the structure so the center of I is in line with the needle structure. Glue it so the slot runs left to right. This will make it easier to insert the knotted portion of your balloon while also holding it securely in place **(figs. ee and ff)**.

FIG. dd

FIG. ee

FIG. ff

### Put the Machine Together

1. We have all the pieces of our machine! But we still have a few more things to do. First we need to attach these two pieces to a table or counter. Use painter's tape to secure it. Line up the edge of the worm gear mechanism structure with the edge of the table, and make sure your paddle/regulator can spin without bumping the table. Secure the hammer mechanism about 12" (30.5 cm) from the corner **(fig. gg)**.

2. Next, we need to find some weights so we can turn our worm drive. I used a bag full of marbles. Marbles are good because you can add them or take them away in order to get the best weight. The amount that worked best for me was 19 marbles. Wrap the rubber band that's attached to the end of the string around the bag of marbles. You can wind up your worm gear by rotating the regulator *counterclockwise*. The wire loop should keep it centered on the cardboard you wrapped around the pencil shaft. If you let go of the paddle, the bag of marbles will fall and the worm will turn. To keep it locked in place, move the lever under the end of the paddle, and then stick a light weight on top of the end of the lever. I just used a strip of cardboard. You want it to be heavy enough to keep the lever from moving and the paddle turning, but light enough to be knocked off by a domino.

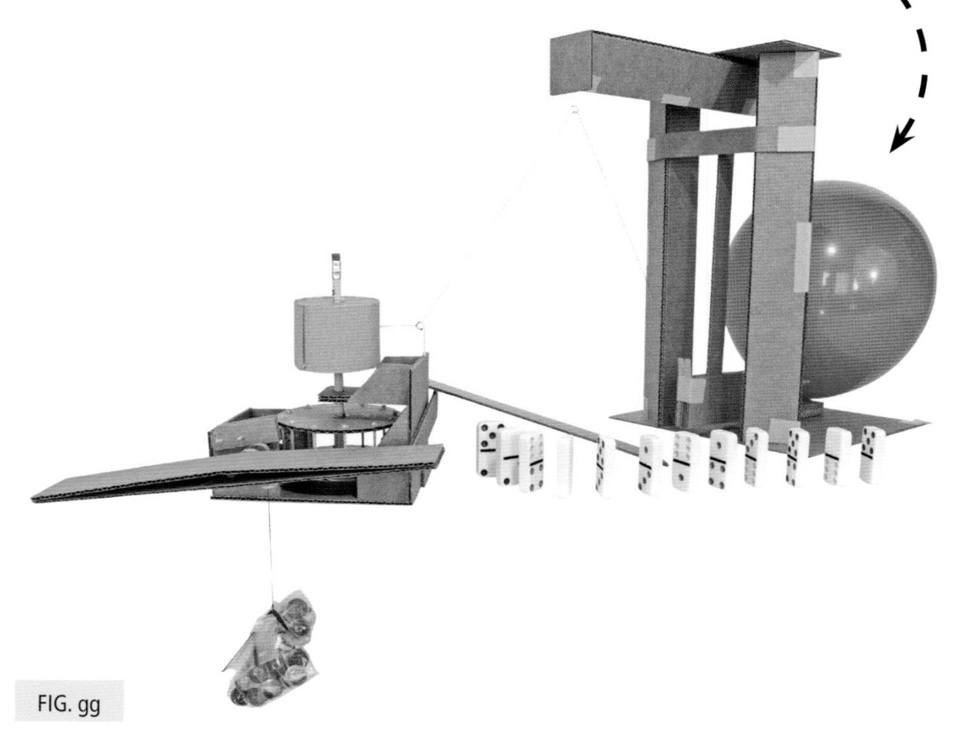
FIG. gg

# START THE CHAIN REACTION!

**The Setup**

1. Inflate your balloon and insert the knot underneath the slotted holder. Make sure the surface of the balloon is resting against the cardboard edge of the popping mechanism (the edge where the needle will come out).
2. Find the string that's attached to the cylinder and run it through both guides and then down to the slit in the thin piece of cardboard attached to your hammer. You want to push the string into the slit, kind of like flossing your teeth. Trim the other side so only about 1" (2.5 cm) of string is sticking out.
3. Now add a row of dominoes so the last one knocks off the cardboard strip resting on the end of the lever.
4. Unless you have a ton of extra balloons, I'd remove the balloon before you try your first test run.

You can't expect something this crazy and complicated to work on the first try. It took me five times before I could get the whole thing to run through smoothly. It might take you less; it might take you more.

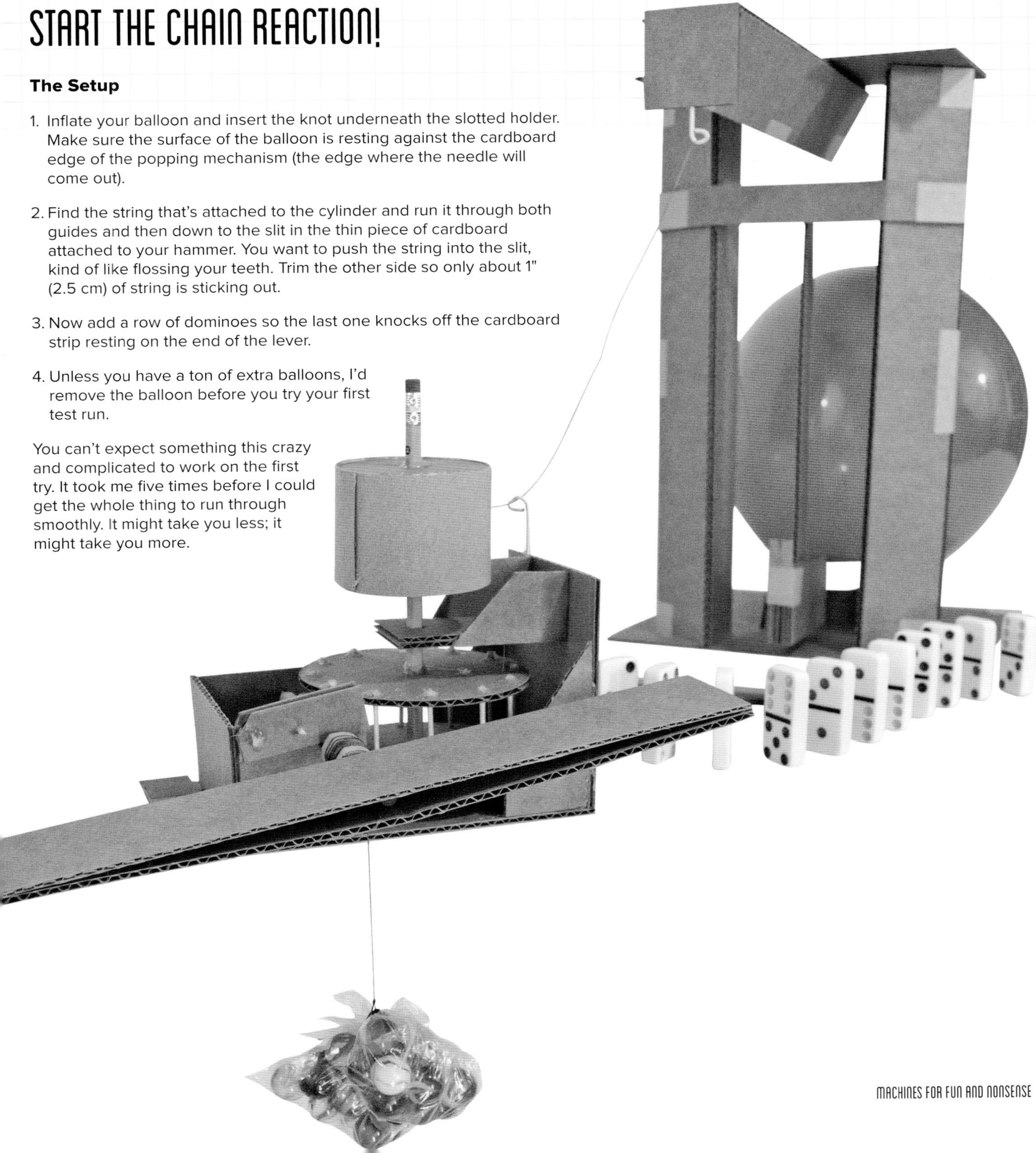

FIG. a

## Make It Go!

1. Knocking over the first domino causes the whole row to fall **(fig. a)**.
2. The last domino knocks off the strip of cardboard, releasing the lever **(fig. b)**.
3. The lever releases the paddle of the regulator. The bag of marbles falls, causing the worm drive to rotate **(fig. c)**.
4. The rotation of the worm drive rotates the lantern gear. The shaft attached to the lantern gear rotates the cylinder. The cylinder winds up the string, which is attached to the hammer **(fig. d)**.
5. The hammer is lifted until it makes contact with the cardboard stopper **(fig. e)**.
6. The string continues to wind, pulling itself out of the slit.
7. The hammer is now released and comes crashing down into the popping mechanism. This forces the needle through the hole in the cardboard, making contact with the balloon, therefore popping it **(fig. f)**!

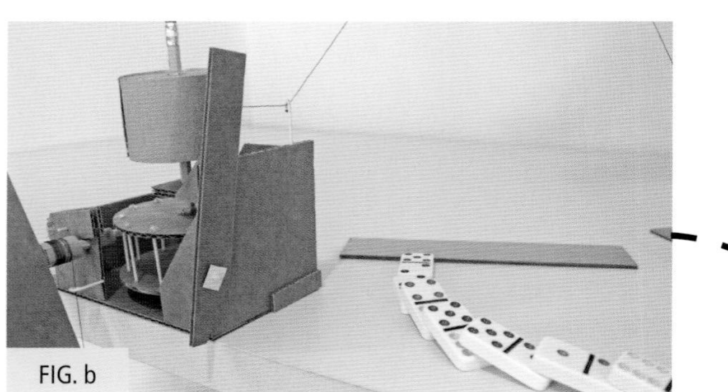
FIG. b

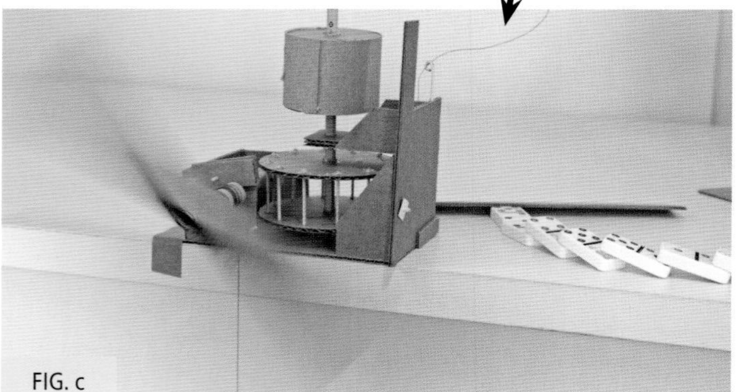
FIG. c

FIG. d

FIG. e

## Engineering Know-How

The regulator, or paddle, keeps the weight from falling too quickly. It does this by making use of air resistance. The surface of the paddle is slowed down by the air. The larger the surface area of the paddle, the slower it's able to turn.

FIG. f

## Troubleshooting

The lantern gear assembly can be a bit tricky to get right. I even had the tip of my pencil break at one point. You can cover the sharpened tip with a single layer of masking tape to give it some structure. Other tapes will probably be too tacky and cause friction.

Depending on the plastic you use for the worm gear, you might get frustrated with it coming undone on the ends. If glue doesn't hold it, you can poke a tiny hole in the ends to loop string through (using a sewing needle makes it easy). Then you can essentially tie the ends to the pencil shaft. Make sure you poke the holes near the very end or else it might still pop loose.

CHAPTER IIIII

# MACHINES FOR FOOD

Sure, you could dunk your own cookies by hand, but there's no fun and engineering in that. The machines in this chapter refer to food, but you could just as easily vend out trinkets and marbles if you don't have candy on hand.

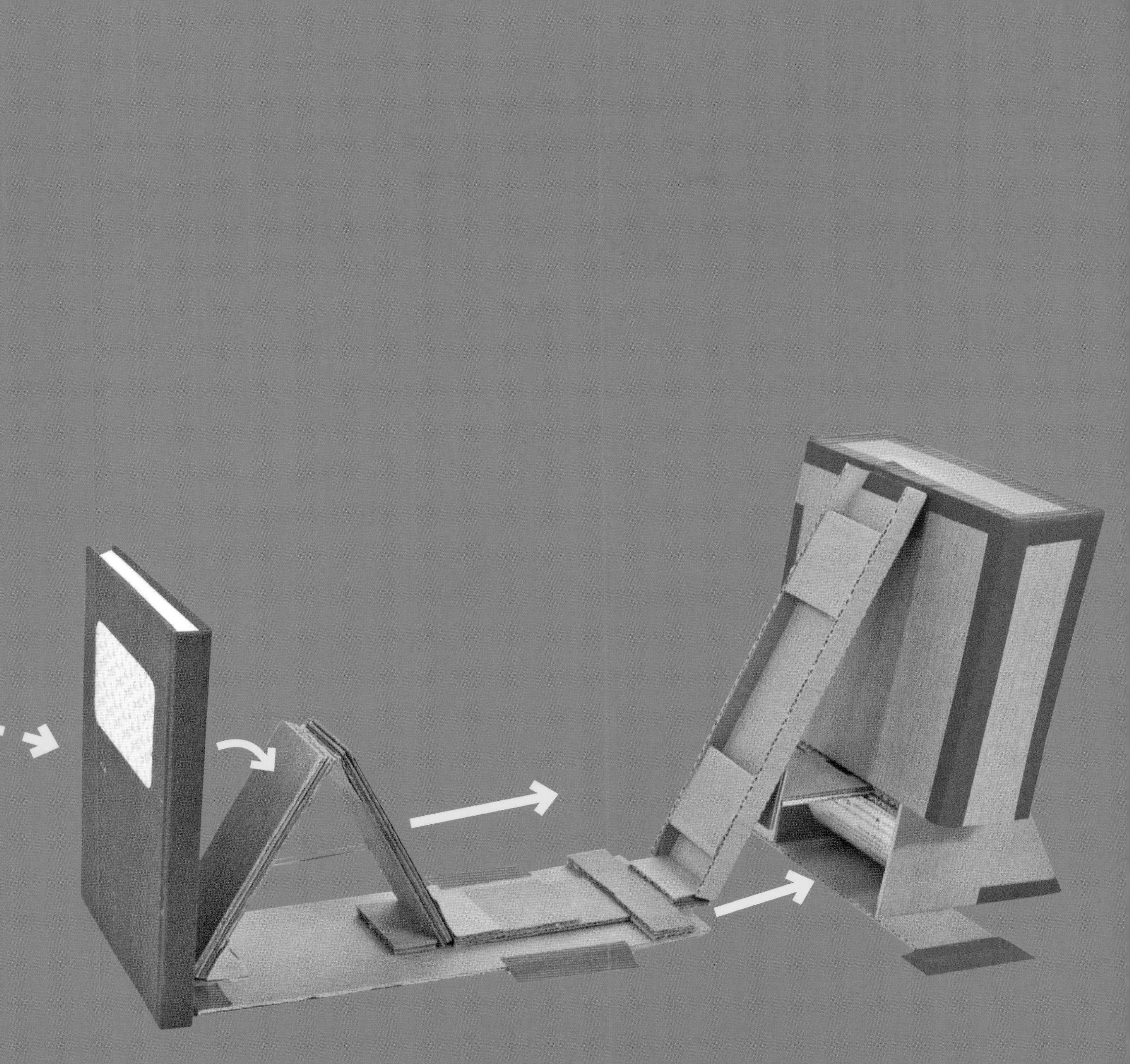

You can use the Candy Dispenser (see page 136) to dole out candy, morsels, or bolts.

# VENDING MACHINE

Who knew a coat hanger and a bit of cardboard could make something as cool and handy as a vending machine? Build a few of these, mount them in a box, stock them with trinkets you're trying to get rid of, and start a college fund with all the money you'll make.

# BUILD MATERIALS & TOOLS

## SPIRAL SCREW

### CARDBOARD:

Four 1" x 1" (2.5 x 2.5 cm)–**A1, A2, A3 & A4**
Four 1½" x 1½" (3.8 x 3.8 cm)–**B1, B2, B3 & B4**
Four 3" x 6½" (7.6 x 16.5 cm)–**C1, C2, C3 & C4**
One 3" x 8" (7.6 x 20.3 cm)–**D**
Two 4" x 9" (10.2 x 22.8 cm)–**E1 & E2**

### OTHER:

Wire coat hanger–**F**
6" (15 cm) craft sticks–**G**
New pencil–**H**

## HARD CYLINDER TO ACT AS MANDREL

Hard sturdy cylinder smaller than 2" (5 cm) in diameter, such as a thin glass jar, wooden dowel, or thick and sturdy cardboard tube **(I)**. This will be used to wrap the coat hanger wire around to make the vending screw, so it needs to be very study and round.

## GEARS

### CARDBOARD:

Two circles, 2" (5 cm) in diameter, with holes poked in the center–**J1 & J2**
Two circles, 6" (15.2 cm) in diameter–**K1 & K2**
Two 1" x 20" (2.5 x 50.8 cm) to make gear teeth–**L1 & L2**

### OTHER:

3 toothpicks–**M**
6" (15 cm) skewer–**N**

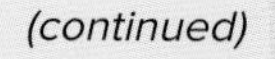
*(continued)*

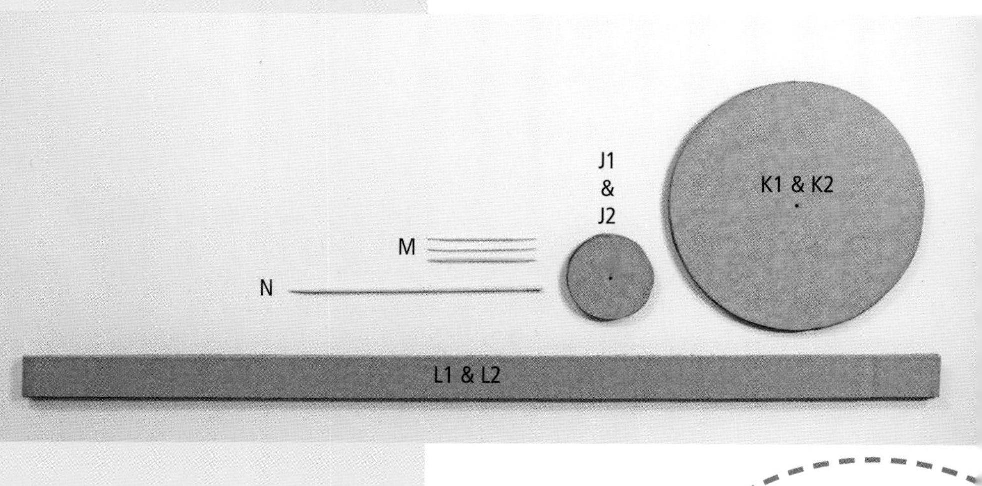

## WEIGHTED LEVER AND TAPE RAMP

**CARDBOARD:**

Three 1" x 10" (2.5 x 25.4 cm)–**O1, O2 & O3**

Two 1" x 4" (2.5 x 10 cm)–**P1 & P2**

One 2" x 5" (5 x 12.7 cm)–**Q**

One 2" x 3" (5 x 7.6 cm)–**R**

One 2" x 2" (5 x 5 cm)–**S**

One 2½" x 12" (6.4 x 30.5 cm)–**T**

**OTHER:**

2 batteries, or some other weight–**U**

## MARBLE RAMP

**CARDBOARD:**

One 1" x 3" (2.5 x 10 cm) thin cardboard–**V**

One 2" x 8" (5 x 20.3 cm)–**W**

One ⅝" x 2" (1.6 x 5 cm)–**X**

Two 2" x 6" (5 x 15.2 cm)–**Y1 & Y2**

**OTHER:**

6" (15 cm) craft stick–**Z**

## NOT SHOWN

Flat candy (like a candy bar) or other flat objects to vend

Roll of tape

Marble

## TOOLS

Rulers

Sharpened pencil

Scissors

Awl

Hot glue gun and glue sticks

Wire cutters or bolt cutters

Pliers

Masking or electrical tape

Leather or rubber gloves (optional)

Eye protection

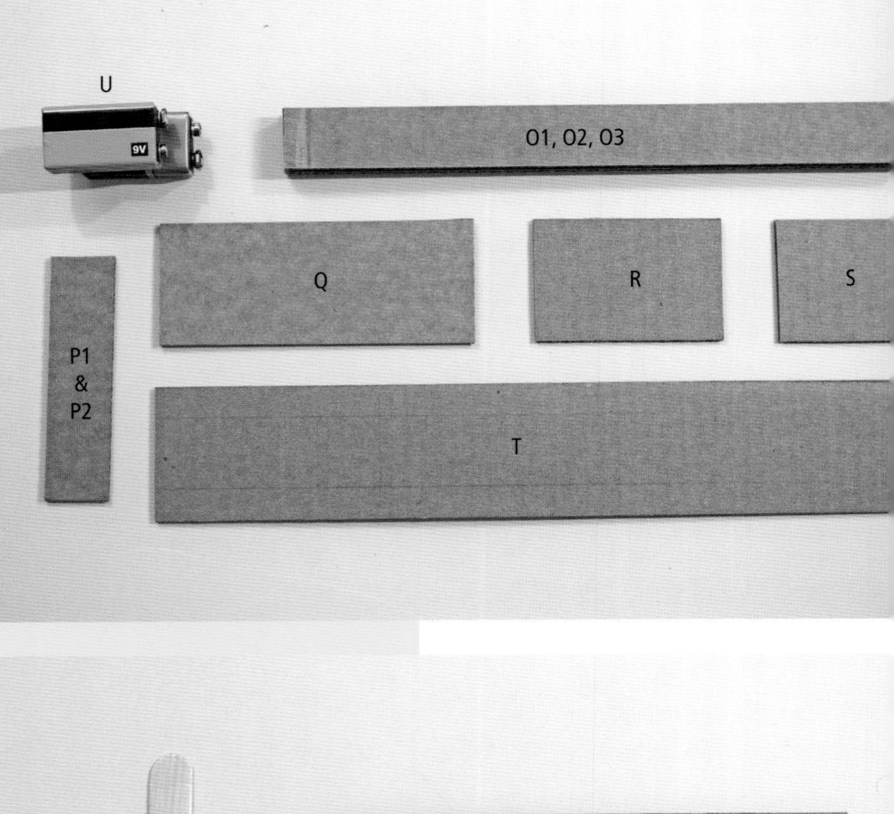

# BUILD THE MACHINE

### Make the Main Structure

1. Measure, mark, and cut a 2" x 4" (5 x 10.2 cm) triangle off of E1 and E2. These will be the sidewalls of the main vending structure **(fig. a)**.

2. Stack C1, C2, C3 and C4. Draw a 3" x 3" (7.6 x 7.6 cm) square on one side and connect the corners with a line. Use an awl to poke a hole where the lines intersect. Make sure all of the edges are lined up before poking the hole; it needs to be in the same spot for all pieces. Glue together the pieces to create two sets. Enlarge the poked hole with a sharp pencil **(fig. b)**.

3. Lay out D, which will be the base, and all of the pieces you prepared in steps 1 and 2 for the main structure. Glue the sidewalls to the base, and then glue the pieces with the poked holes. Attach one piece even with the back edge and glue the other 2" (5 cm) from that. Pay attention to the orientation of the poked holes. They should both be at the bottom **(figs. c and d)**.

4. Create a gear using J1 and J2 and the corrugated strip (see page 18). Slip a new pencil through the holes **(fig. e)**.

5. Add A1 and A2 as spacers between the gear and the end of the pencil and insert it into the holes in the structure. Push the pencil so about ½" (1.3 cm) is exposed on the inside of the structure. Secure the gear to the pencil with glue, but make sure you don't accidentally attach it to the main structure **(fig. f)**.

6. Poke a hole in the center of all four B pieces. Slide B1 onto the pencil until it's tight against the upright structure. Glue B2, B3, and B4 together and slide them onto the pencil. It's best to add a little glue to the pencil and then slide them on. The goal is to have the first piece act as a spacer and have the other three pieces be attached to the pencil. If you rotate the gear, the pencil plus the three pieces should all rotate **(fig. g)**.

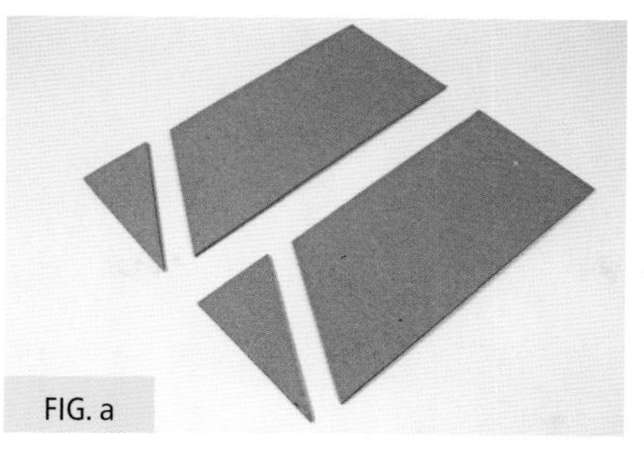
FIG. a

FIG. b

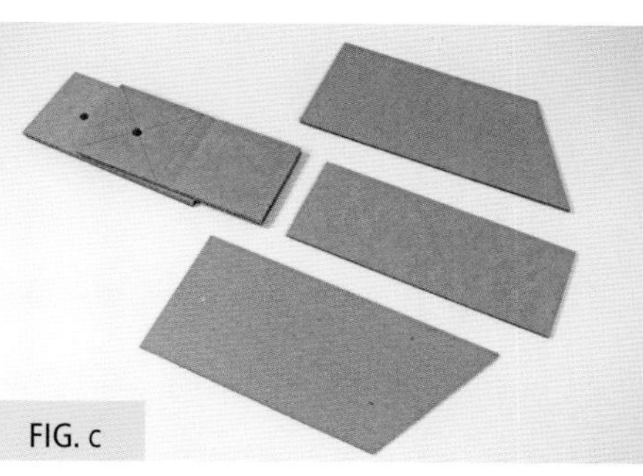
FIG. c

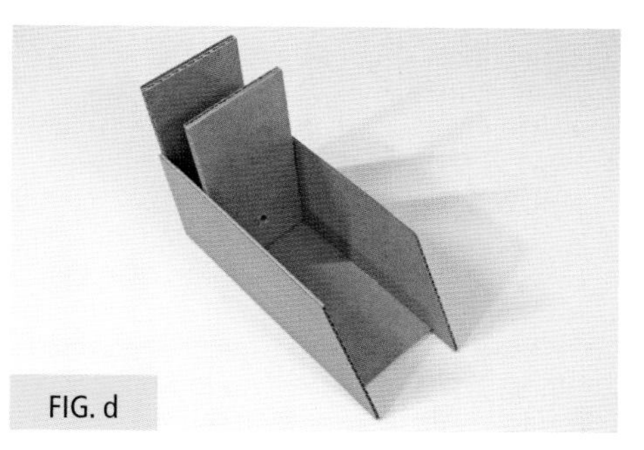
FIG. d

FIG. e

FIG. f

FIG. g

## Make the Vending Screw

1. With wire cutters or bolt cutters, cut the hook part off of the coat hanger and use pliers to straighten it. If you're using a glass cylinder, add tape around it to give it some texture and protect it. If you have thick cardboard or wood, tape won't be necessary. Add masking or electrical tape to the ends of your hanger so the sharp edges don't scratch you **(fig. h)**.

2. Firmly hold one end of the wire against the cylinder. If you have leather or rubber gloves you should wear them. Slowly and carefully wrap the other end around the cylinder. (CAUTION: As you get closer to the end, the wire will want to rebound if you let go. It's recommended you wear eye protection in case this happens. Just use common sense, go slow, and if it's too difficult, try to find a thinner coat hanger. In my experience, ones with a white coating are thinner than the others.)

   Note that I wrapped my wire around the cylinder in a *clockwise* direction. If you happened to go counterclockwise, it's no big deal. It'll just change the way you attach the weighted lever later on **(fig. i)**.

3. Depending on the size of your cylinder, your wire should have been able to loop around three or four times. Remove the wire from the cylinder and carefully bend and pull on your spring until it's evenly spaced **(fig. j)**.

4. Once everything is even, look down the center of the spring. Use pliers to create two bends at the end of the wire, with one bend that's in the exact center of the circle **(fig. k)**.

5. Line up the center bend with the pencil. Double-check that your spring fits inside of the structure and can rotate freely without getting caught on the sidewalls. If your spring is crooked at the end and is sticking out and catching, gently bend or squeeze it until it fits nicely. Once the fit is good, attach it with glue and cover with tape to help it stay put **(fig. l)**.

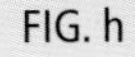

FIG. h

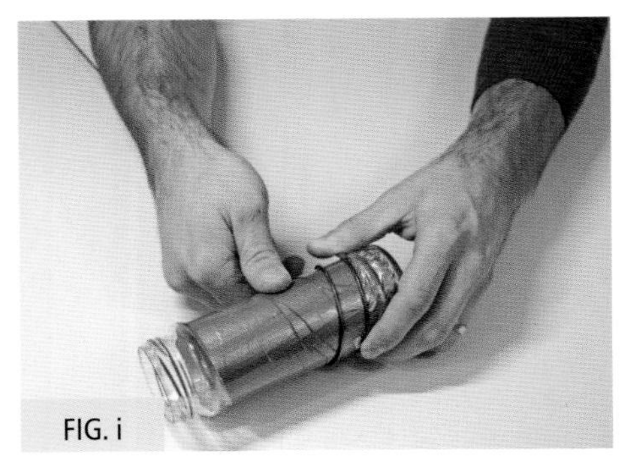

FIG. i

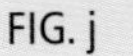

FIG. j

FIG. k

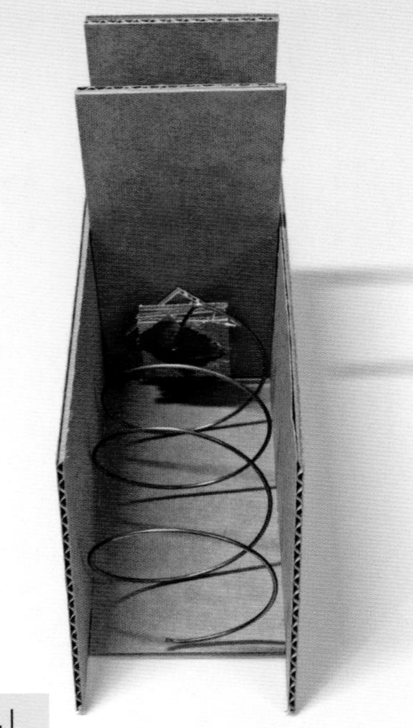

FIG. l

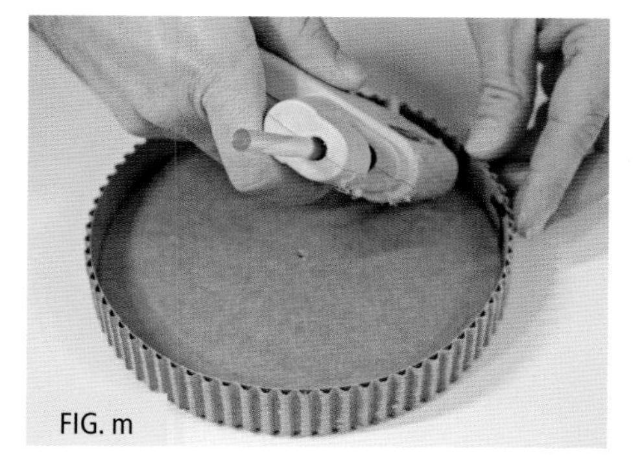

FIG. m

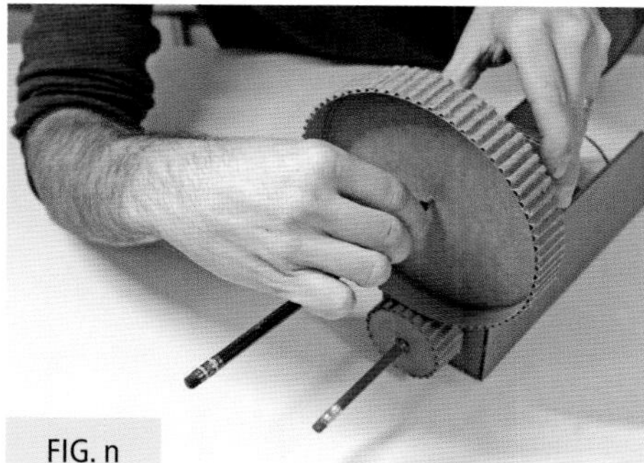

FIG. n

FIG. o

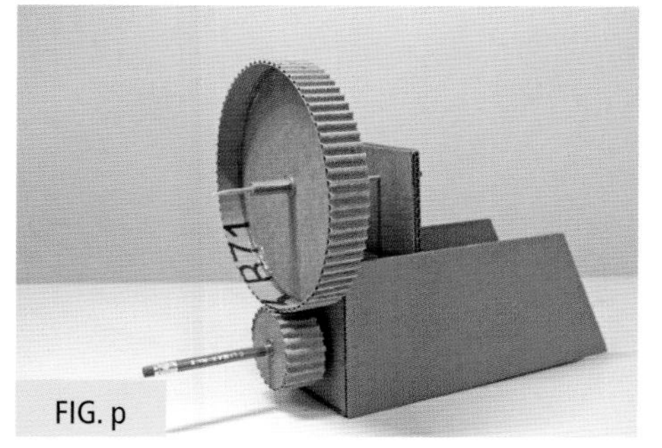

FIG. p

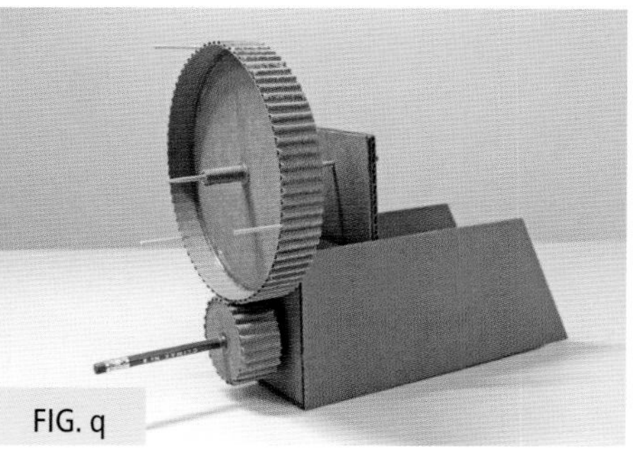

FIG. q

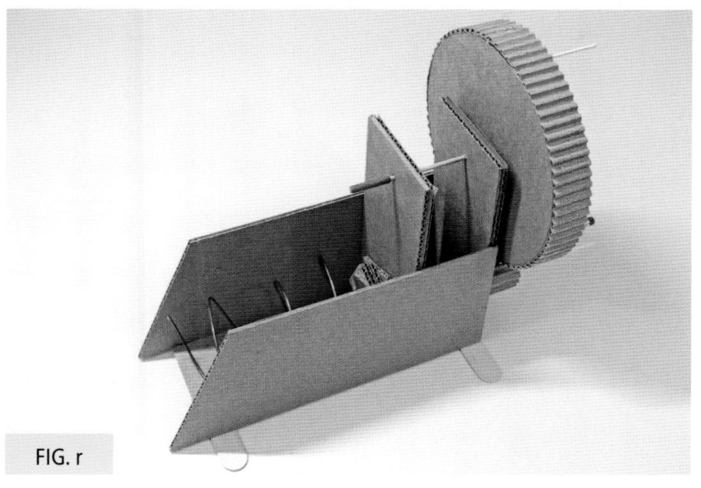

FIG. r

### Make the Large Gear

1. Remove the top layer of cardboard from L1 and L2 to reveal the corrugated side. Find the center of K1 and K2 (see page 12) and poke a hole in the center. Double up K1 and K2 and add the strip with exposed corrugation to make a gear (see page 18). For this gear, we can leave the outer circle off. Add a bead of glue around the inner edge **(fig. m)**.

2. Set the 6" (15.2 cm) gear (K) on top of the 2" (5 cm) gear (J) and center it on the main structure. Make sure the teeth are meshed properly and then insert a pencil into the center hole of the gear and make a mark on the cardboard behind the gear. Remove the gear and poke a small hole in the center of the mark **(fig. n)**.

3. Use a skewer to extend the hole to the other piece. Make sure the skewer is level and square before poking the hole in the other piece of cardboard. Add tape around the skewer on the inside and slide on A3 and A4 on the outside to act as washers **(fig. o)**.

4. Slide on the gear, then roll a strip of thin cardboard (or the paper you removed from the cardboard strips to create the gears) around the skewer. Add glue or tape to keep it from unfurling, and push it tight against the gear. Make sure it's perpendicular to the gear, and then add a healthy amount of glue. The key is not to have any wobble when the gear rotates. If you need some extra support, use little square pieces of cardboard to attach the base of the gear to the rolled-up piece **(fig. p)**.

5. Count up all of your gear teeth and divide by three. (Keep track by writing the number on the corresponding tooth for every 10 you count.) I had 75 teeth, but you might have a different number. Thus, 75 ÷ 3 = 25, so I slid a toothpick into the gear every 25th tooth. Don't attach them with glue just yet.

   If you have 76, you could go 25, 25, 26, or for 74 go 24, 25, 25. The goal is to divide it up into three equal sections the best you can (see page 135).

6. Glue two craft sticks to the bottom of the main base **(fig. r)**. This is so we can secure it to a sturdy surface with tape later on.

### Make the Weighted Lever

1. Remove the top layer of cardboard on P1 to expose the corrugation. Lay it next to the other materials for the weighted lever **(fig. s)**.
2. Glue the corrugated piece of P1 to one O piece **(fig. t)**.
3. Glue the three O pieces together with the piece that has P1 attached on top. Then tape the batteries (or other weights) to the end **(fig. u)**.
4. Cut a slight angle in P2 **(fig. v)**.

### Make the Tape Ramp

1. Score T ½" (1.3 cm) from the edges (see page 14). Score a line 3½" (9 cm) from the end and then cut and remove the 3½" x ½" (9 x 1.3 cm) pieces from the corners **(figs. w and x)**.
2. Glue S to the bottom edge of the tapered end of the ramp, and then add R underneath to lock it in place. Glue the base (Q) to the bottom **(fig. y)**.

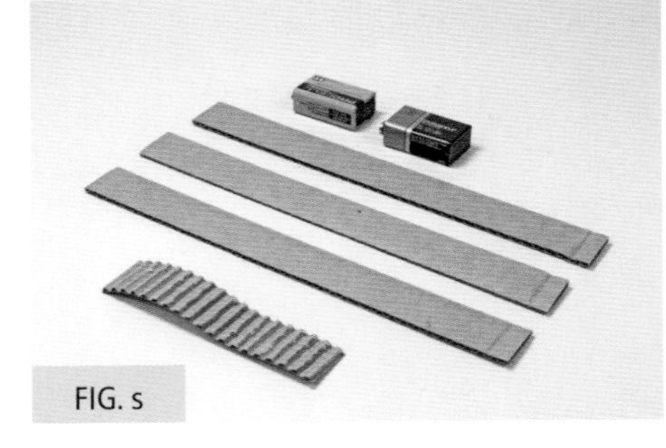
FIG. s

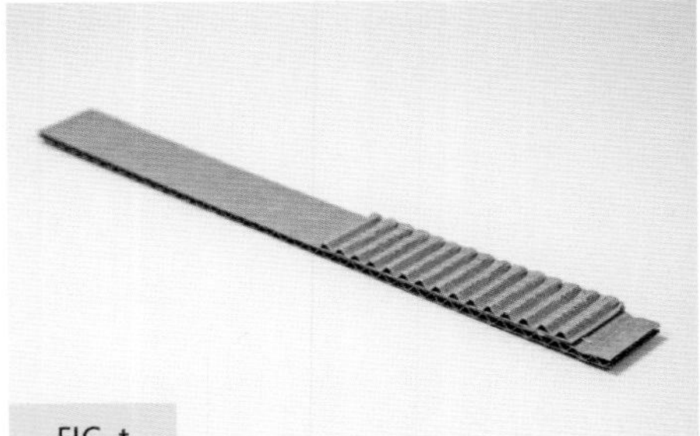
FIG. t

FIG. u

FIG. v

FIG. w

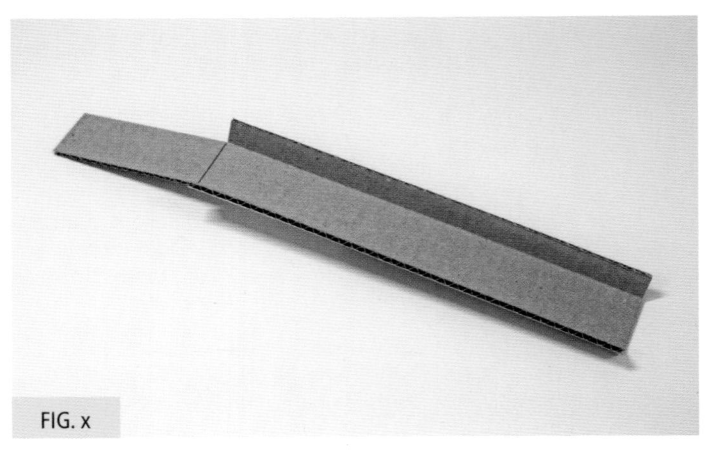
FIG. x

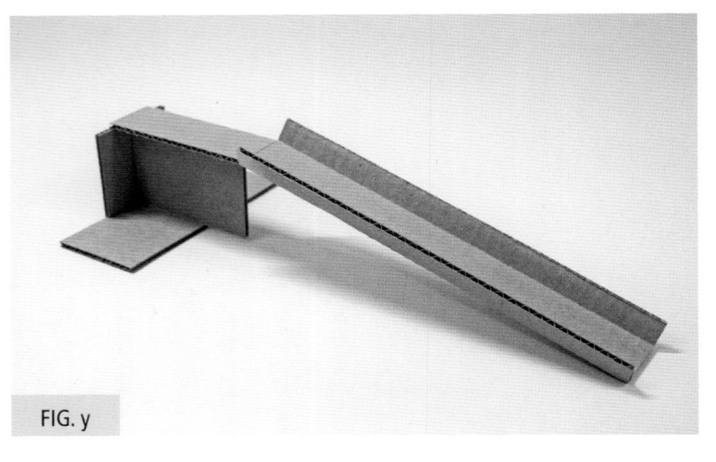
FIG. y

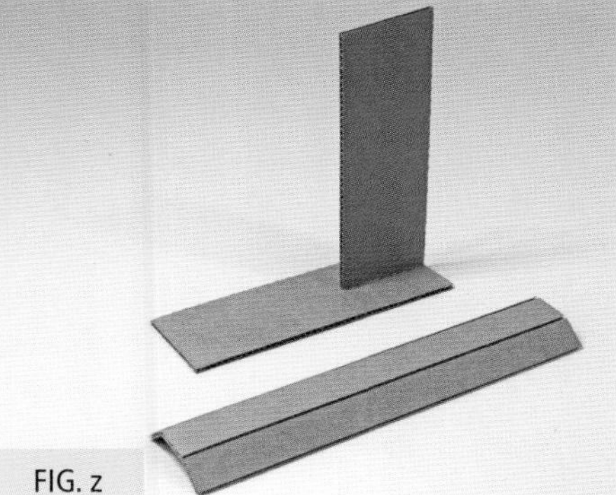

FIG. z

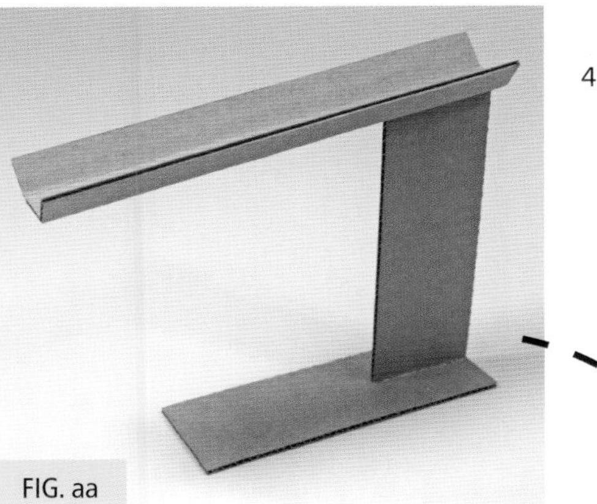

FIG. aa

FIG. bb

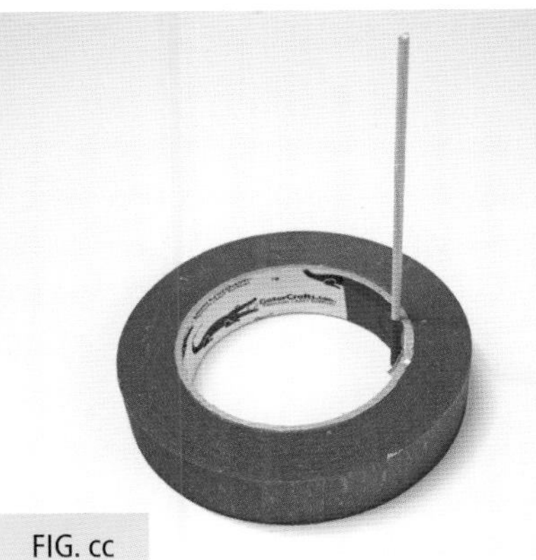

FIG. cc

## Make the Marble Ramp

1. Lay out the pieces for the marble ramp and craft stick lever. Score W ⅝" (1.6 cm) from the edges (see page 14). Cut a slight angle into Y1, and glue the flat edge perpendicular to Y2 **(fig. z)**.
2. Attach the scored marble ramp channel to the angled piece **(fig. aa)**.
3. Glue X to the bottom of the craft stick, 1½" (3.8 cm) from the end, in a T-formation. Create a U-shape with V and glue it to the other end of the craft stick as shown **(fig. bb)**.
4. Tape a skewer to the inside of a roll of 1" (2.5 cm) wide tape **(fig. cc)**.

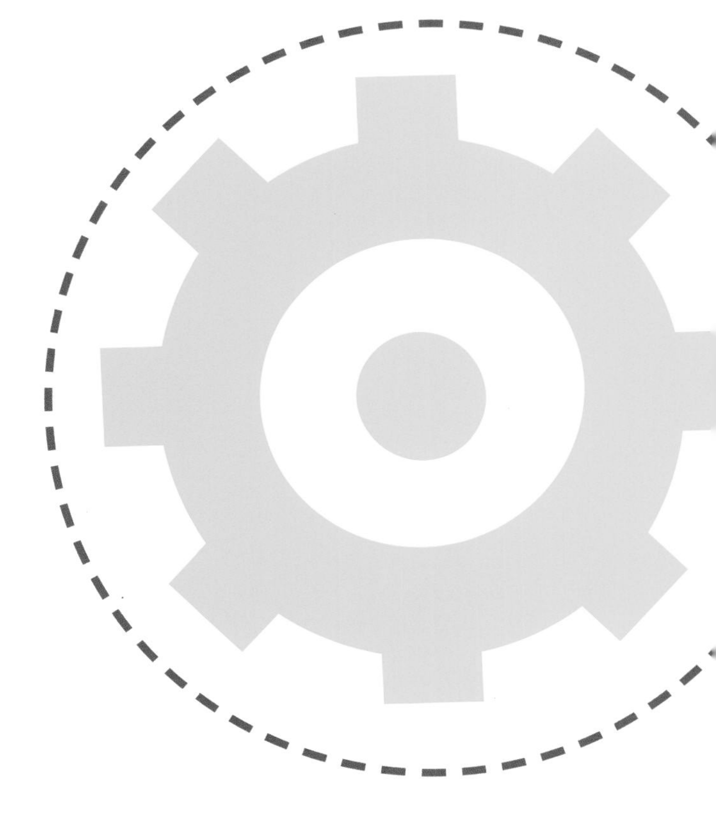

# START THE CHAIN REACTION!

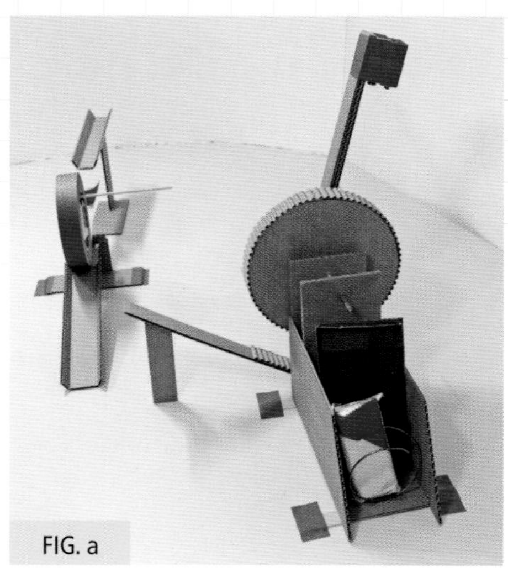
FIG. a

### The Setup

1. Tape the vending machine structure to a tabletop to secure it. Find candy or items that will fit into the vending screw. I used a candy bar and a pack of cookies **(fig. a)**.

2. Position the 6" (15.2 cm) gear so one of the toothpicks is at approximately 11:00. (If you wound your wire counterclockwise, you would make it so the toothpick is at 1:00.) While holding onto the large gear, let the weighted lever rest against this toothpick. Take the strip of cardboard with the corrugation glued to the end and wedge it on the underside of the 2" (5 cm) gear on the right side. (Again, if you did a counterclockwise wind, you would put it on the left side.) The teeth should mesh together, which will keep the gear from turning. Place the 4" (10.2 cm) piece with the angled cut underneath the end of this piece to hold it in place **(fig. b)**.

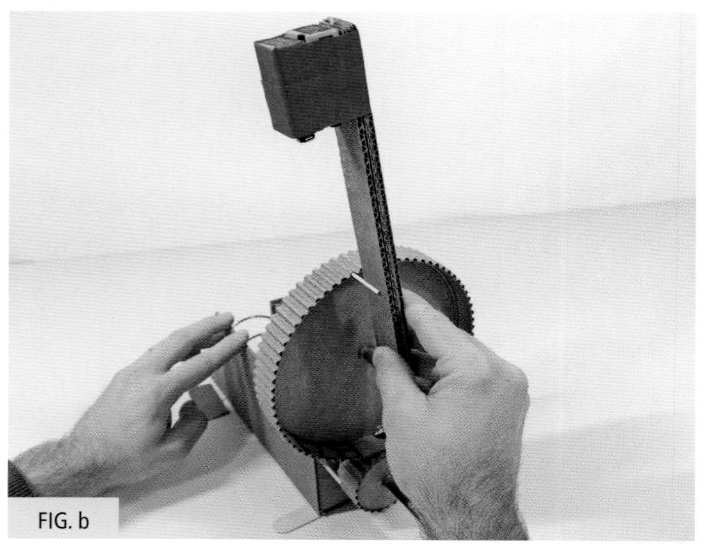
FIG. b

3. Line up the tape ramp so that the skewer attached to the roll of tape will bump the 4" (10.2 cm) angled piece when it rolls past it. Secure it to the tabletop with tape. Carefully balance the roll of tape on top of the flat platform. If it wants to roll off, place a thin sheet of paper in front of it. Set the craft lever on the back edge. It should be sensitive enough that when a marble falls into the U shape, it will cause the tape to start its forward roll.

4. Line up the marble ramp so the end is right above the U-shaped piece **(fig. c)**.

FIG. c

### Make It Go!

To set off the chain reaction, let a marble roll down the first ramp. When it falls into the U-shaped end of the craft lever, it should start the tape rolling forward. If the ⅝" x 2" (1.6 x 5 cm) piece of cardboard was in the middle of the craft stick, you would need a heavier weight to make this happen. However, because it's off-center and closer to the roll of tape, we can use a small weight, like a marble, to supply enough force to get the tape rolling. By placing the pivot point close to the roll of tape, we're creating mechanical advantage. The longer the distance from the U-shaped end to the pivot point, the less weight we need to get the tape rolling.

When the tape rolls down the ramp, the skewer that's sticking out will bump into the angled piece holding the corrugated lever in place. When the corrugated lever falls, it allows the 2" (5 cm) gear to move. The weighted lever, which is already slightly angled to the left, will now fall, spinning the 6" (15.2 cm) gear one-third of a turn. This in turn rotates the 2" (5 cm) gear one full rotation. Because the vending screw is attached to the 2" (5 cm) gear via the pencil, it too rotates one turn, causing the item (in this case, a small package of cookies) to fall out the front.

To reset the machine, you don't need to rotate the vending screw. Simply slide off the weight lever, rotate it, and then position it back in place so it's resting against the second toothpick (which should be at the 11:00 position now).

### Engineering Know-How

Earlier, we divided the 6" (15.2 cm) gear into three sections. Why did we do that? Turns out, it wasn't an arbitrary number. The 6" (15.2 cm) gear is three times larger than the 2" (5 cm) gear (2" x 3" = 6"). Not only does that mean the diameter is three times larger, but it also means the 6" (15.2 cm) gear has three times as many teeth as the 2" (5 cm) gear. So, when the 2" (5 cm) gear rotates one full turn, the 6" (15.2 cm) gear only rotates one-third of a turn. If we had used an 8" (20.3 cm) and a 2" (5 cm) gear, we would have divided the larger gear into four sections, because one-fourth of a turn of the 8" (20.3 cm) gear would give us a full rotation in the 2" (5 cm) gear. If we had used anything smaller than 6" (15.2 cm), like a 4" (10.2 cm) gear, for example, we would need to rotate it half a turn to get one full rotation with the 2" (5 cm) gear. The issue there is that the table would get in the way of the falling lever, and it wouldn't be able to rotate a half turn. We would need to hang it over the edge of the table so it could get the half rotation (or 180 degrees) that we would need. The only catch with using a larger gear, like a 6" (15.2 cm) or even an 8" (20.3 cm) gear, is that we need more weight on the end of the lever to provide enough force to rotate the 2" (5 cm) gear.

### Troubleshooting

Any time you make your own gears, you're likely to have some issues because they have to be pretty precise to work correctly. If your circles aren't perfectly round, or your corrugated teeth are a bit smashed, you'll run into issues. If you have any glue on the spots where the gears mesh together, you'll get a bit too much friction, which could keep the gears from spinning smoothly.

You also might have an issue with your vending screw scraping the edge of the sidewalls, or not being centered around the rotating pencil. Just be patient here and gently bend and squeeze and test and retest. The wire is very malleable, so you should be able to dial it in eventually.

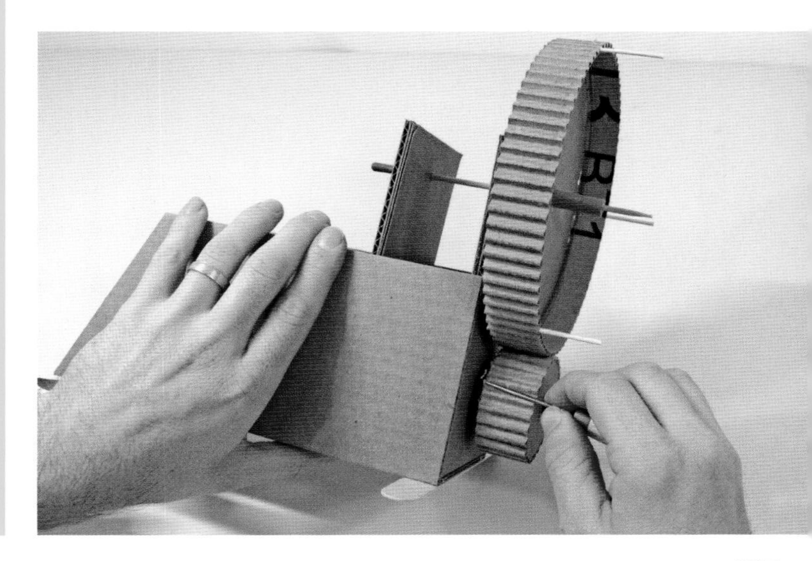

# CANDY DISPENSER

"Candy" is in the name, but you could just as easily dispense marbles or tiny nuts and bolts. Why you would rather have nuts and bolts over candy is beyond me, but to each their own.

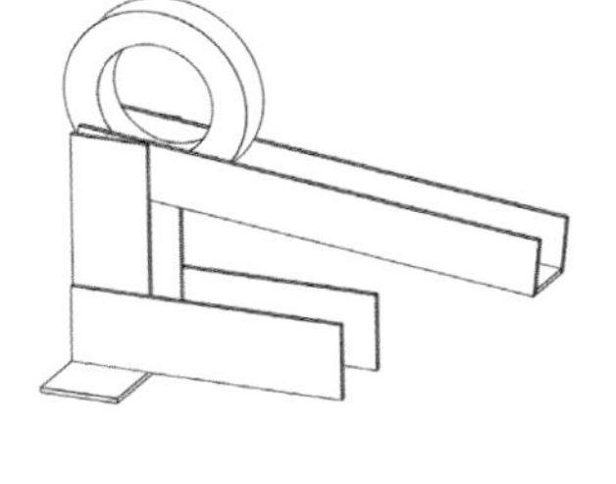

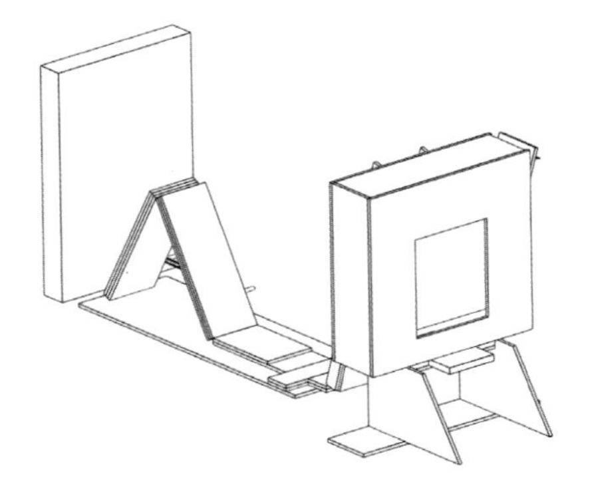

# BUILD MATERIALS & TOOLS

### RAMP

**CARDBOARD:**

Three 2" x 6" (5 x 15.2 cm)–**A1, A2 & A3**
Two 2" x 7" (5 x 17.8 cm)–**B1 & B2**
One 1¼" x 12" (3.2 x 30.5 cm)–**C**
Two 1½" x 12" (3.8 x 30.5 cm)–**D1 & D2**

**OTHER:**

6" x 9" (15.2 x 22.8 cm) book or block of wood–**E**

### TRIANGULAR SLIDER

**CARDBOARD:**

One 1" x 4" (2.5 x 10.2 cm)–**F**
One 2" x 3" (5 x 7.6 cm)–**G**
One 2" x 6½" (5 x 16.5 cm)–**H**
Two 1" x 1" (2.5 x 2.5 cm)–**I1 & I2**
Four 1" x 2" (2.5 x 5 cm)–**J1, J2, J3 & J4**
Six 2" x 5½" (5 x 11.4 cm)–**K1, K2, K3, K4, K5 & K6**
One 4" x 12" (10 x 30.5 cm)–**L**

**OTHER:**

Rubber band–**M**
2 toothpicks–**N**

### CANDY MACHINE BODY

**CARDBOARD:**

Two 2" x 5" (5 x 12.7 cm)–**O1 & O2**
Two 2" x 2" (5 x 5 cm)–**P1 & P2**
Five 2" x 7" (5 x 17.8 cm)–**Q1, Q2, Q3, Q4 & Q5**
One 1½" x 2" (3.8 x 5 cm)–**R**
One 3" x 10" (7.6 x 25.4 cm)–**S**
Two 7" x 7" (17.8 x 17.8 cm)–**T1 & T2**
One 2" x 4" (5 x 10.2 cm)–**U**
One 2" x 8" (5 x 20.3 cm)–**V**
Two ¾" x 1½" (1.8 x 3.8 cm) thin cardboard–**W1 & W2**
Four 1" x 2" (2.5 x 5 cm)–**X1, X2, X3 & X4**
Two 3" x 5" (7.6 x 12.7 cm) with 2" x 3" (5 x 7.6 cm) triangular cutout–**Y1 & Y2**
One 4" x 7" (10.2 x 17.8 cm) thin cardboard–**Z**

**OTHER:**

4" x 5" (10.2 x 12.7 cm) thin clear plastic–**AA**
Rubber band–**BB**

### NOT SHOWN

Small candy like M&Ms

### TOOLS

Hot glue gun and glue sticks
Utility knife
Straightedge
Painter's tape
Awl
Sharpened pencil

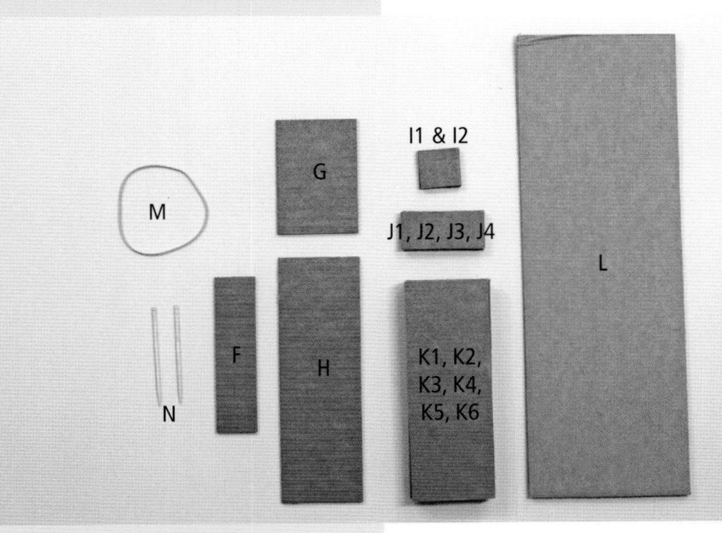


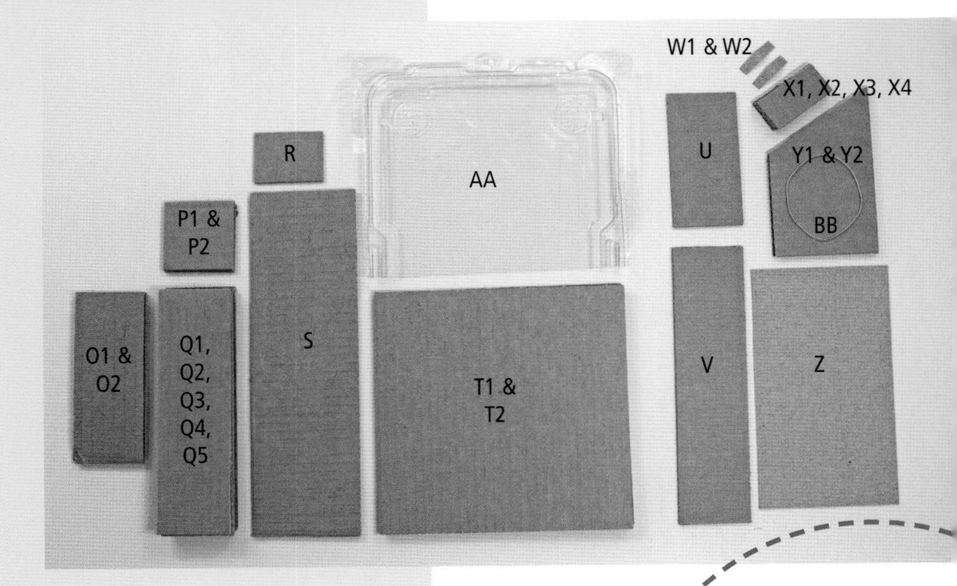

# BUILD THE MACHINE

### Make the Ramp

1. Lay out C, D1, and D2. Glue the two outer pieces to the center piece to create a channel **(fig. a)**.
2. Add A1 and A2 to the ends at a slight angle to create a slightly sloped ramp **(fig. b)**.
3. Add B1 and B2 to create feet, and then glue A3 to the rear for extra balance **(figs. c and d)**.

### Make the Triangular Slider

1. Glue together two sets of the K pieces, each one three layers thick **(fig. e)**.
2. Glue J1, J2, J3, J4, and G to H as shown in **fig. f**.

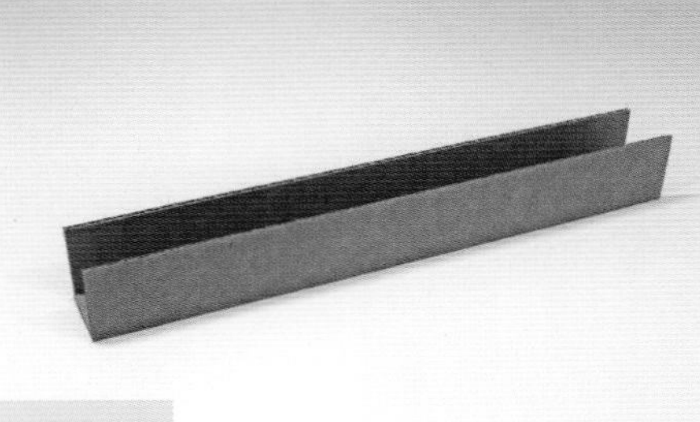

FIG. a

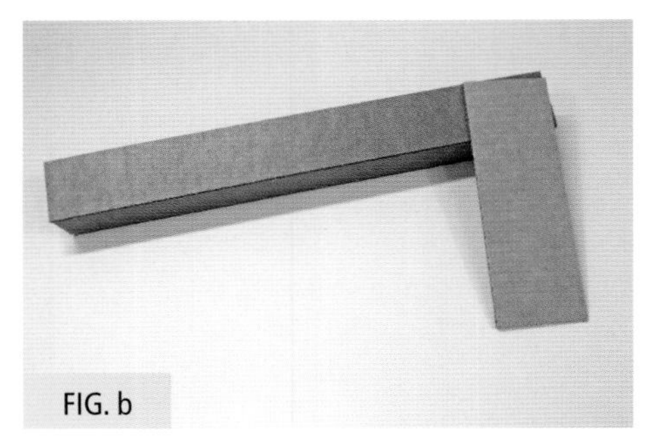
FIG. b

FIG. c

FIG. d

FIG. e

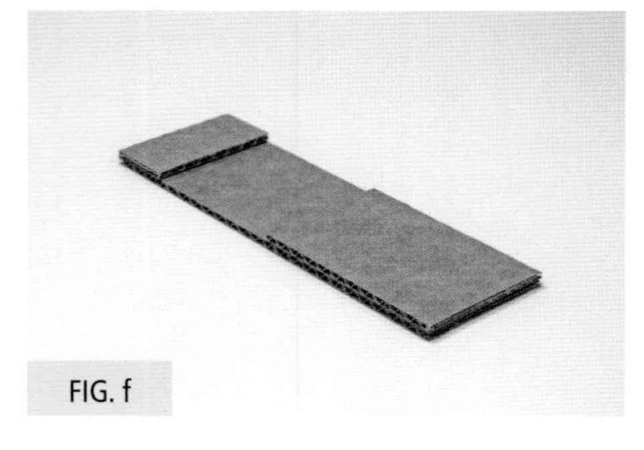
FIG. f

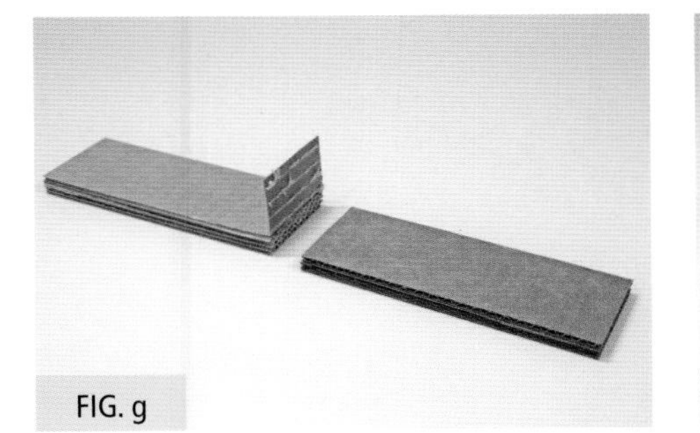
FIG. g

FIG. h

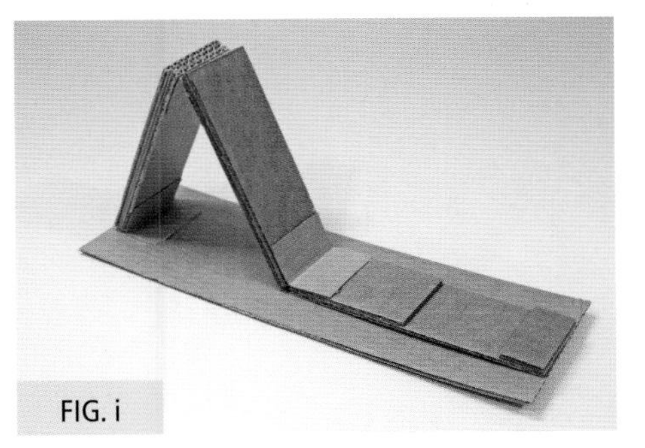
FIG. i

FIG. j

FIG. k

FIG. l

3. Create hinges between A stacks and B1. Tape probably won't be strong enough, so it's best to use cardboard skin for the hinge. Use glue to attach it **(figs. g and h)**.
4. Attach this whole piece to the base (L) with a hinge on the end of the A stack **(fig. i)**.
5. Glue I1 and I2 spacers to F. Place this over the single layer section of the flat piece as shown. Make sure the flat piece can slide back and forth. This small piece is meant to keep the sliding piece flat but still allow forward and backward movement. If it doesn't seem to move freely, you can add a thin spacer to the I pieces to give it a little extra height. Once you're sure it can move freely, glue it to the base, making sure you don't accidentally glue the sliding piece **(figs. j and k)**.
6. Insert the toothpicks near the base of the triangular uprights. If they slide around, add a little glue. Wrap the rubber band around them. If you gently push on top of the triangular tip, it should flatten out slightly and the slider should move forward. When you let go, it should pop back into place **(fig. l)**.

## Make the Candy Dispenser

1. With a utility knife and straightedge, cut a window in T1, 1" (2.5 cm) from the bottom and 2" (5 cm) from all other sides **(fig. m)**.

2. Cut out a window from the thin clear plastic that's 1" (2.5 cm) or so bigger than your cutout. Glue it to the cardboard and secure the edges with tape **(fig. n)**.

3. The size of your candy and how much you want to be dispensed each time will dictate what size hole to cut out of this next piece. For chocolate M&Ms, this size should be perfect. For anything larger, you'll have to experiment with enlarging the cutout **(fig. o)**.

   Take Q1 and Q2 and mark a 1" x 1" (2.5 x 2.5 cm) square in the center. Glue the pieces together and cut out where you marked. If it's easier, you can cut them first and then glue them together. This will be the bottom of your candy dispenser.

4. Score a line 2" (5 cm) across Q3 (see page 14). Add a tape handle to help you open and close it. This is the hatch where you add candy **(figs. p and q)**.

5. Lay out T2, Q4, and Q5 and all of the pieces you have prepared in steps 1–4. It doesn't matter which side the hatch is on, but the double-layered piece with the cutout should be on the bottom. **(fig. r)**

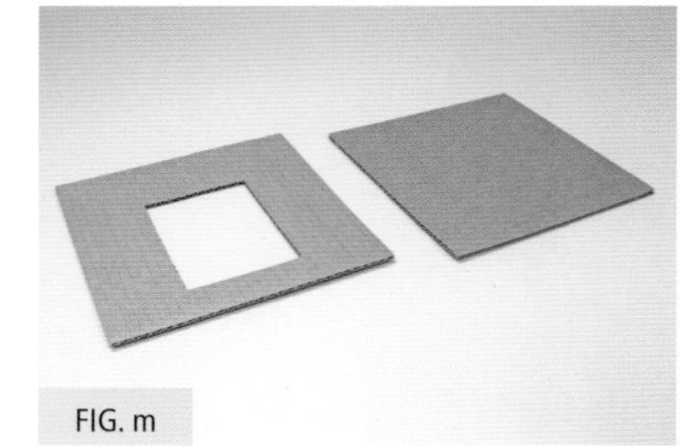

FIG. m

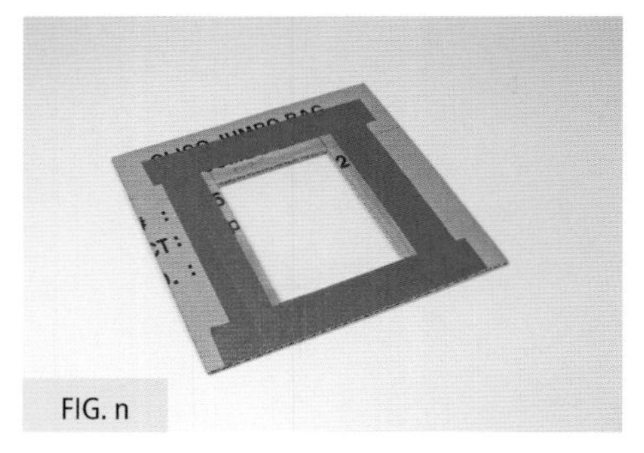

FIG. n

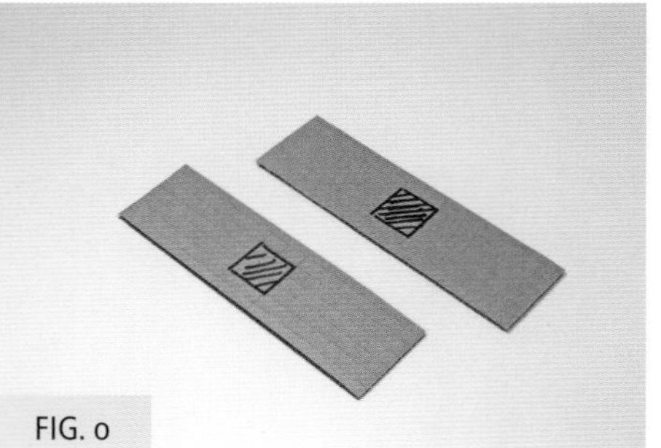

FIG. o

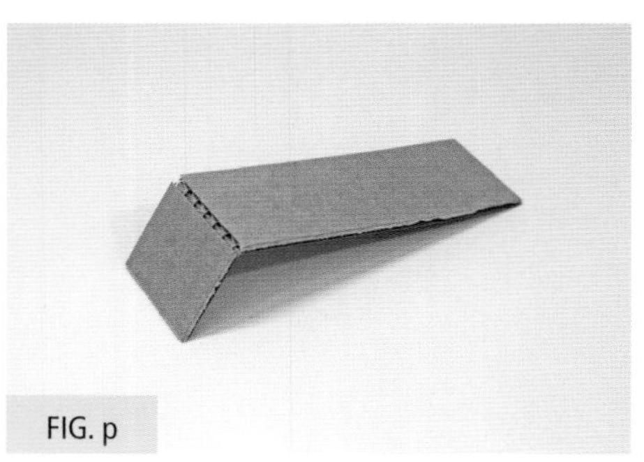

FIG. p

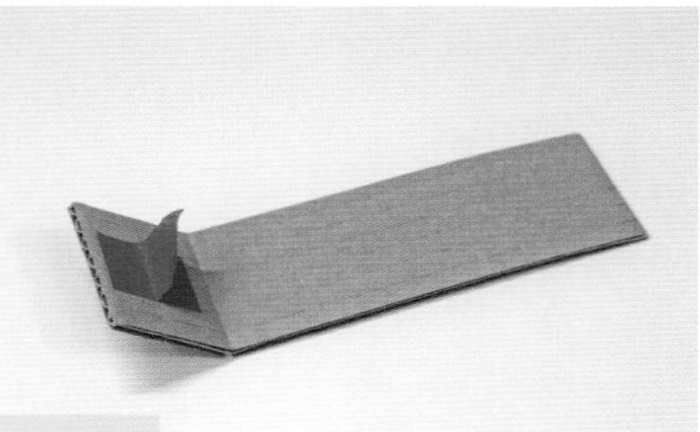

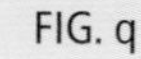

FIG. q

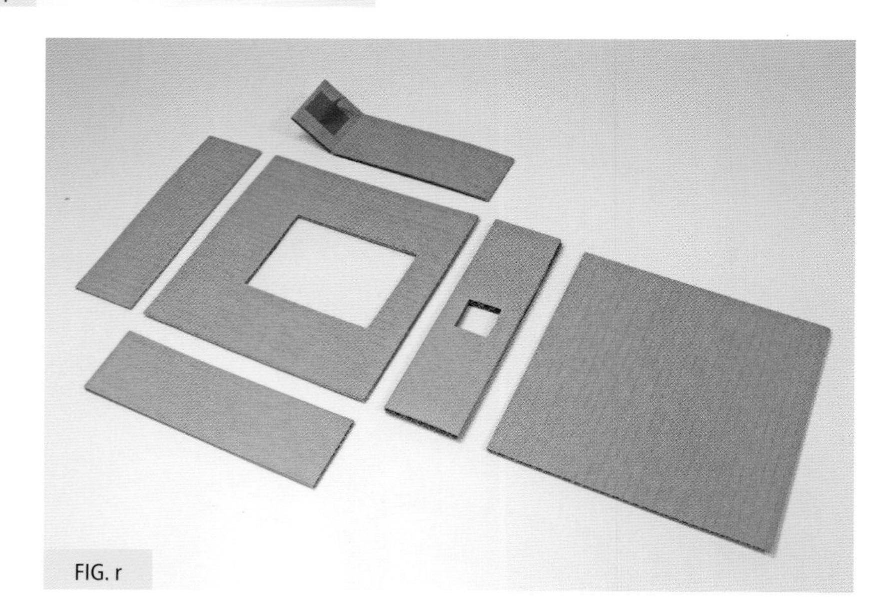

FIG. r

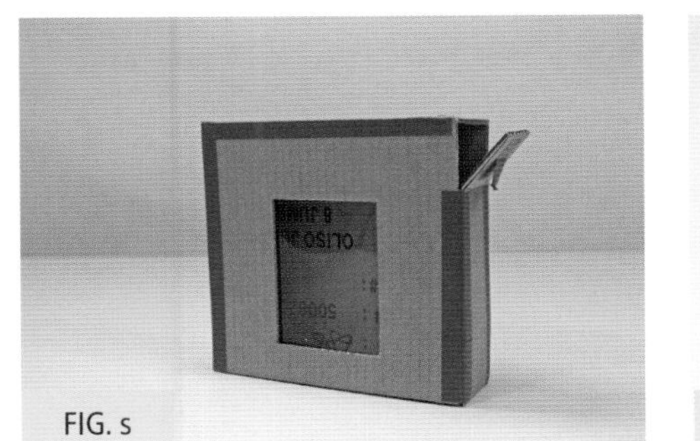
FIG. s

FIG. t

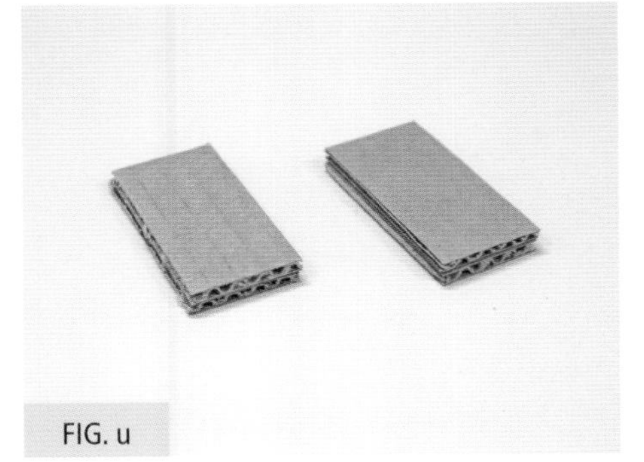
FIG. u

FIG. v

FIG. w

6. Glue all of the pieces together and add tape to the seams for extra holding power **(fig. s)**.
7. Mark a 1" x 1" (2.5 x 2.5 cm) square in the center of O1 and O2, glue them together, and cut out **(fig. t)**.
8. Glue the X pieces together in two pairs **(fig. u)**.
9. Flip over the candy machine body so the bottom is facing up. Line up the two cutouts and glue down X1, X2, X3, and X4 spacers on either side, making sure the O pieces can slide back and forth **(fig. v)**.
10. To have a little extra room, add scrap pieces of thin cardboard as a spacer **(fig. w)**.

11. Cut a 1" x 1" (2.5 x 2.5 cm) square from the center of U and glue it to the spacers **(fig. x)**.
12. Add Y1 and Y2. The square edge should be flush with the back **(fig. y)**.
13. Glue R to the back of the slider **(fig. z)**.
14. Fold Z ½" (1.3 cm) from both ends but in the opposite directions **(fig. aa)**.
15. Glue one side to the back edge of U. Curve it to the front and tape it to the front edge. This is where the candy will collect **(figs. bb and cc)**.
16. Glue V to the back of the base for stability **(fig. dd)**.

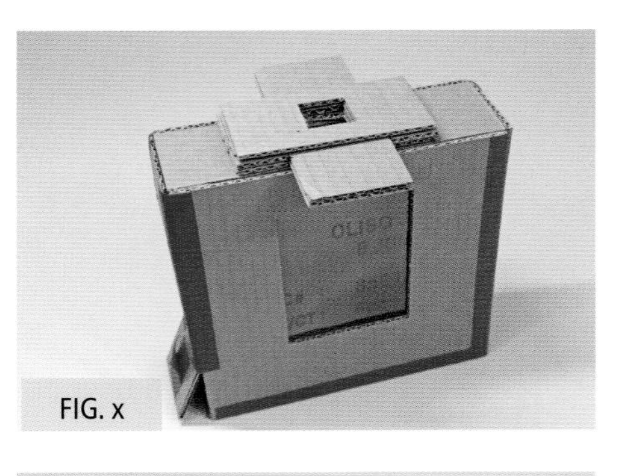

FIG. x

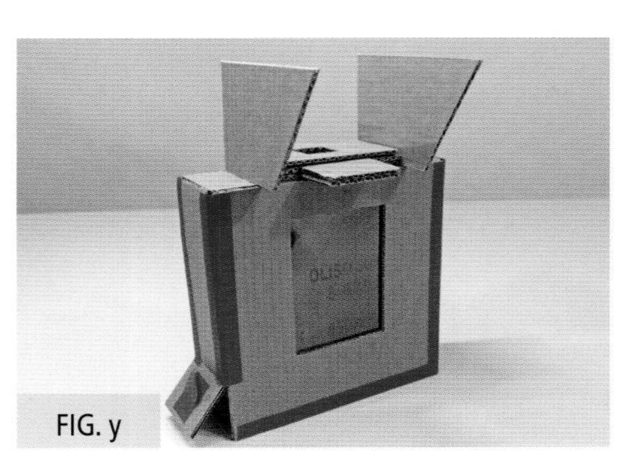

FIG. y

FIG. z

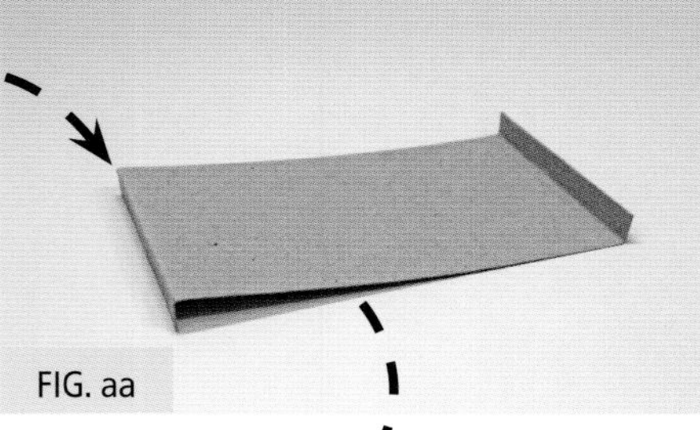

FIG. aa

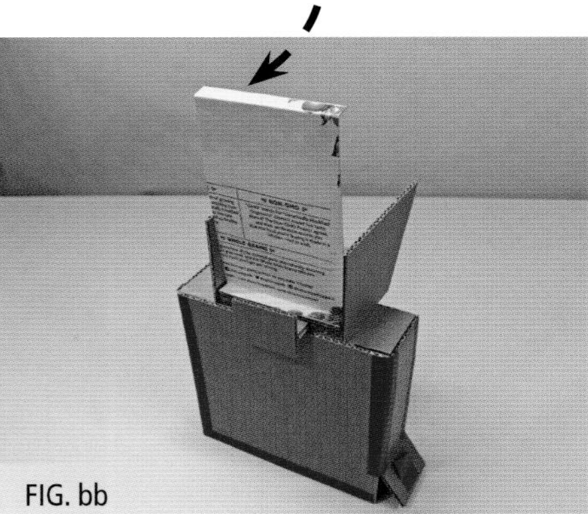

FIG. bb

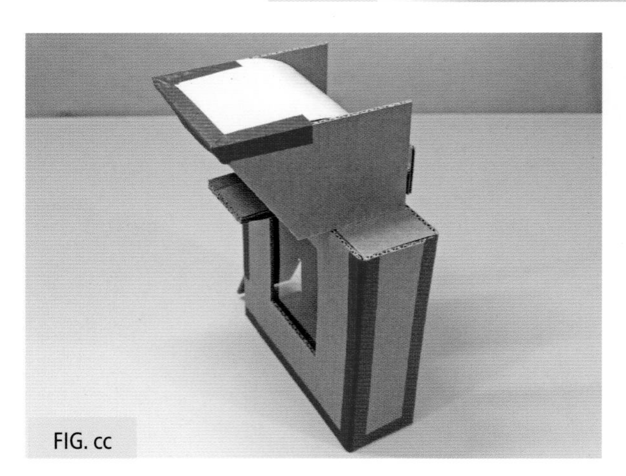

FIG. cc

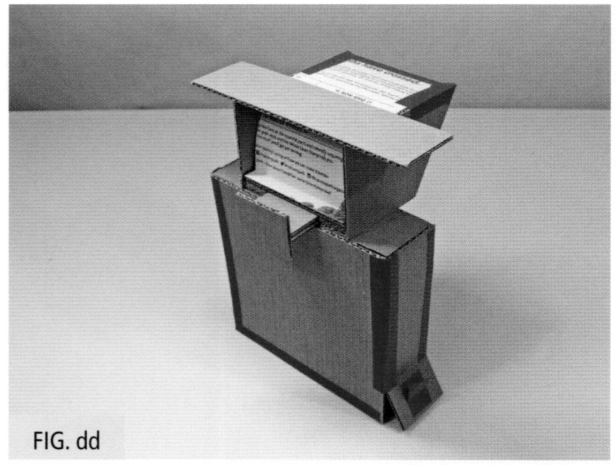

FIG. dd

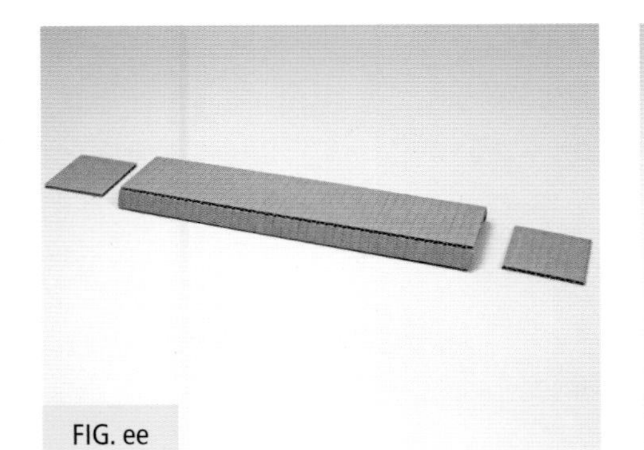
FIG. ee

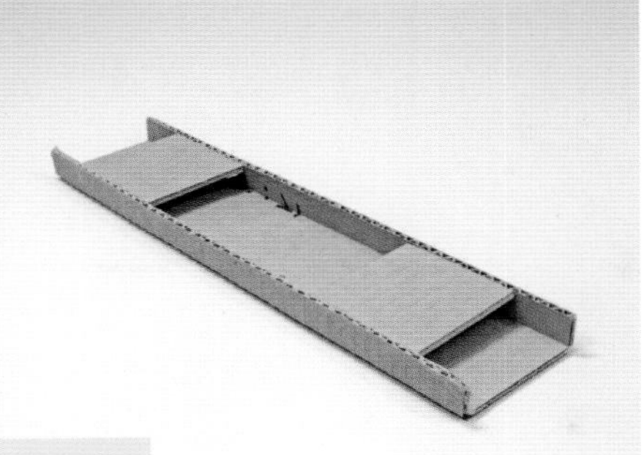
FIG. ff

FIG. gg

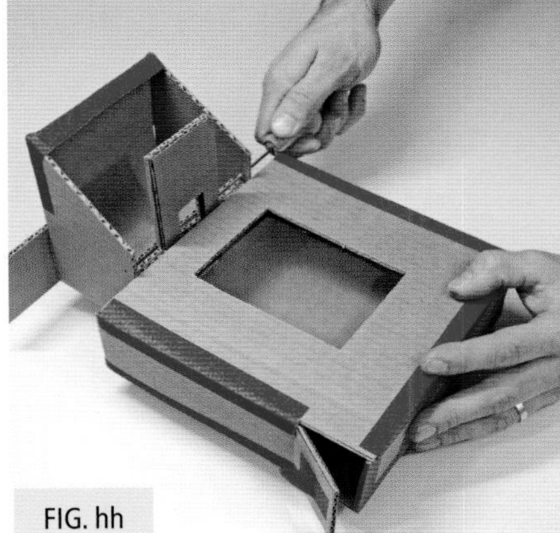
FIG. hh

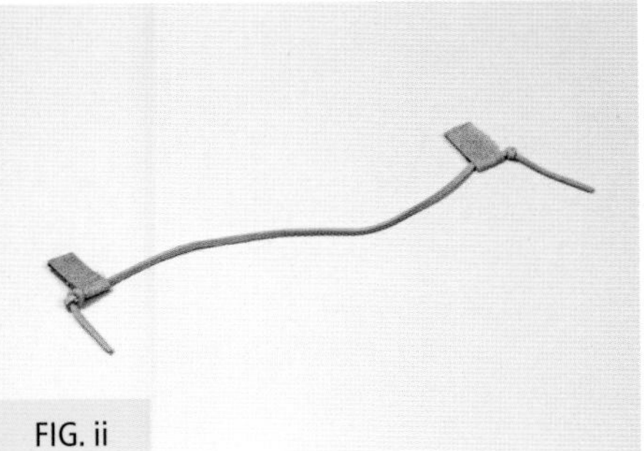
FIG. ii

## Make the Back Lever

1. Score the sides of S lengthwise ½" (1.3 cm) from each edge (see page 14). Fold up to create a channel **(fig. ee)**.
2. Glue P1 and P2 1" (2.5 cm) from the ends to give it structure. Don't glue it to the very ends because we need those to be clear and open **(fig. ff)**.
3. Tape the lever to the back center of the candy dispenser body. Notice that it comes in contact with the flat piece on the slider **(fig. gg)**.
4. With an awl, poke holes on both sides of the candy collector and enlarge with a sharpened pencil. If you made a pokey pad (see page 11), now is a good time to use it **(fig. hh)**.
5. Cut the rubber band so it's one long piece. Tie knots on the end and glue onto W1 and W2 **(fig. ii)**.
6. Slide the flaps through the poked holes and glue them to the sides, making sure the rubber band is slightly tensioned but not too tight **(fig. jj)**.

FIG. jj

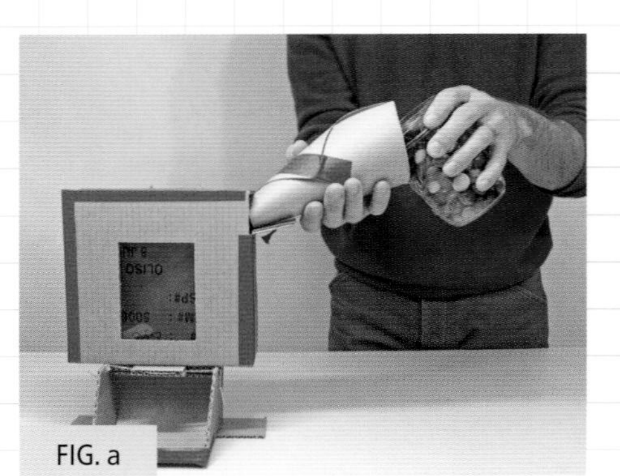

FIG. a

FIG. b

FIG. c

FIG. d

# START THE CHAIN REACTION!

### The Setup

1. Fill the machine with your candy of choice. I'm using peanut butter–filled M&Ms **(fig. a)**.

2. Place the triangular slider so it's touching the lever **(fig. b)**.

3. Place the book or block of wood behind the triangular lever with the right edge of the book in line with the right edge of the triangle. If you center it, the book will fall on the triangular lever and keep it down. If you place it to one side, it will push down on the triangular lever when it falls, but then roll off to the side, allowing the lever to pop back up to its resting position. This is what will open and close the candy release slider inside the dispenser **(fig. c)**.

4. Place the ramp so it's at an angle to the book. If it's head-on, it might prohibit the book from sliding backward or rolling off the triangular lever. Place a roll of tape on the ramp, but don't let it go yet **(fig. d)**.

## Make It Go!

To set off the chain reaction, release the roll of tape. It'll roll down and bump the book. The book will fall on the triangular slider, pushing the horizontal slider mechanism forward. This piece will push on the longer lever, which will push on the dispenser slider. Once it goes forward far enough, the cutouts will line up and the candy will fall through. When the book slides down or falls off the triangle, everything will slide back, which closes off the flow of candy.

## Troubleshooting

The most common problem is that the candy won't come out. That can easily be fixed. Your first option is to find smaller candy. If your heart is set on the larger sugary bits, then you can enlarge all of those 1" x 1" (2.5 x 2.5 cm) cutouts. I found that for normal M&Ms, this was a good size, but it was a little too small for peanut butter M&Ms; they kept getting stuck. You can try to just enlarge the cutout in the slider, but if that doesn't work, you'll need to enlarge all of the holes, which can be a bit tricky without removing the bottom piece. If you're careful, you can stick a knife in there and gently saw away at the edges to make the hole bigger. I made mine 1" x 1½" (2.5 x 3.8 cm) and that seemed to work well.

## Engineering Know-How

The falling book is rotational motion, because its base acts as a pivot. The book swings around that pivot, sort of like the hinge on a door. When it hits the triangular ramp lever, that rotational motion is converted to linear motion. The back-and-forth sliding motion at the bottom is linear, because it goes in a line rather than a circle. That linear motion pushes on the long lever arm, which has a very short rotational motion, because it's hinged on top. That very short rotational motion is then converted back into linear motion when it pushes on the candy dispenser. So we have rotational > linear > rotational > linear all happening in this one tiny machine.

# COOKIE DUNKER

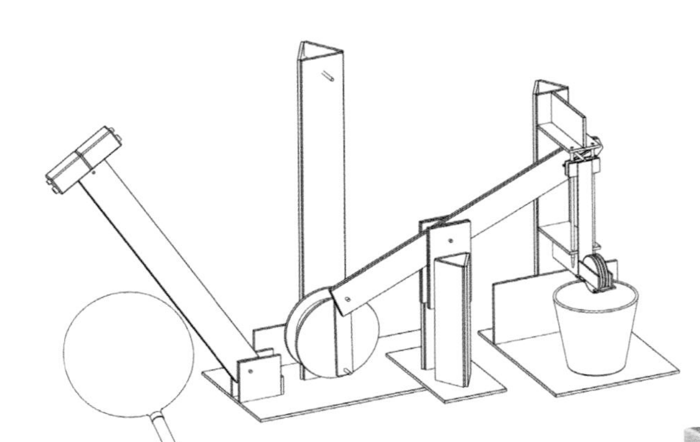

This machine is the perfect excuse to eat cookies. You can tell your parents you need cookies in order to learn about mechanical engineering. And be sure to "troubleshoot" the machine in order to dunk a few extra cookies while you work out the kinks.

# BUILD MATERIALS & TOOLS

## DUNKING MECHANISM

### CARDBOARD:

Two ¾" x 4" (1.8 x 10.2 cm) thin cardboard–**A1 & A2**
Four 1" x 1" (2.5 x 2.5 cm)–**B1, B2, B3 & B4**
Two 1½" x 2¼" (3.8 x 5.7 cm)–**C1 & C2**
Two 1½" x 3½" (3.8 x 9 cm)–**D1 & D2**
One 4½" x 12" (11.4 x 30.5 cm)–**E**
One 3" x 6" (7.6 x 15.2 cm)–**F**
One 6" x 8" (15.2 x 20.3 cm)–**G**

### OTHER:

Four 6" (15.2 cm) skewers–**H**
Four 6" (15.2 cm) craft sticks–**I**

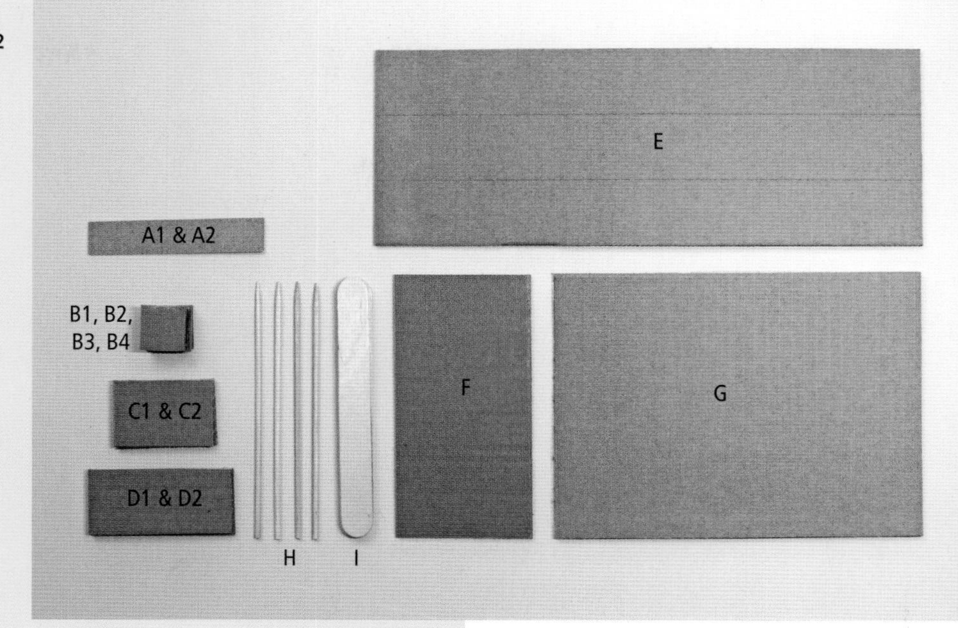


## MAIN STRUCTURE

### CARDBOARD:

One 6" x 16" (15.2 x 40.6 cm)–**J**
One 6" x 11" (15.2 x 28 cm)–**K**
One 2" x 5" (5 x 12.7 cm)–**L**
One 1" x 12" (2.5 x 30.5 cm)–**M**
Two circles, 4" (10.2 cm) in diameter–**N1 & N2**
Two circles, 3" (7.6 cm) in diameter–**O1 & O2**
Two ¾" x 4¼" (1.8 x 10.8 cm)–**P1 & P2**
Two 2" x 2" (5 x 5 cm)–**Q1 & Q2**
One 3" x 5" (7.6 x 12.7 cm) thin cardboard–**R**
Three 1½" x 14" (3.8 x 35.6 cm)–**S1, S2 & S3**

### OTHER:

Balloon–**T**
Two 9V batteries or other weight–**U**
3' (1 m) strong string–**V**
Pencil–**W**

*(continued)*

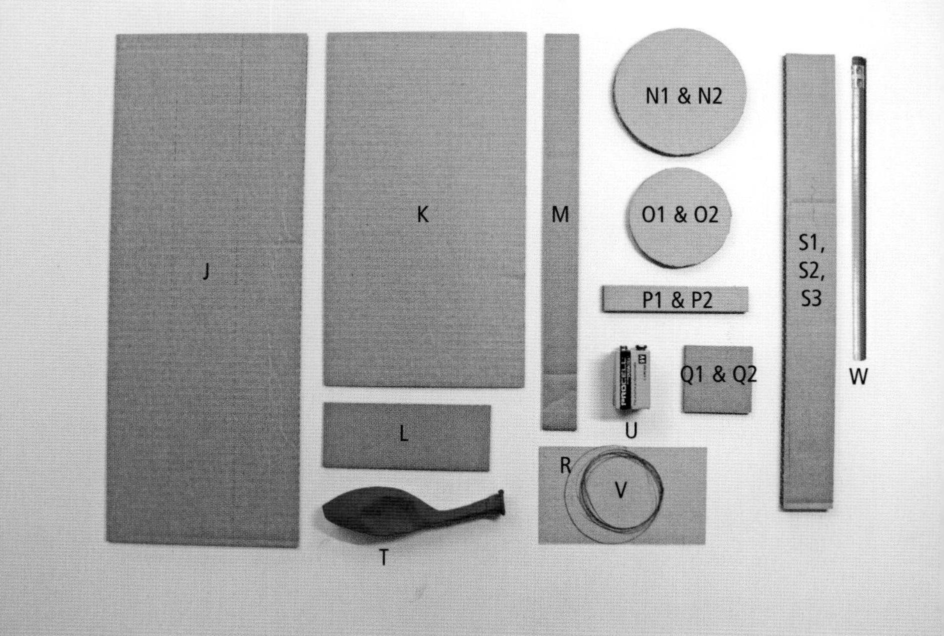

## LEVER TOWER STRUCTURE

**CARDBOARD:**

Two 4½" x 6" (11.4 x 15.2 cm), scored in 1½" (3.8 cm) sections (see page 14)– **X1 & X2**

Four 2" x 4" (5 x 10.2 cm)–**Y1, Y2, Y3 & Y4**

Two 1½" x 14" (3.8 x 35.6 cm)–**Z1 & Z2**

2½" x 5½" (6.4 x 14 cm)–**AA**

**OTHER:**

Cookies–**BB**

Small glass and milk–**CC**

## TOOLS

Utility knife

Hot glue gun and glue sticks

Painter's tape

Awl

Scissors

Sharpened pencil

# BUILD THE MACHINE

### Make the Dunking Mechanism

1. With a utility knife, score E so it's divided into three even sections (see page 14 for how to score). Fold on the score lines to create a triangular tube. Fasten with glue and tape. Glue this to the middle of the base (G) on the long side. Pay attention to the orientation. Add F for support **(figs. a and b)**.

2. Roll A1 and A2 around the skewers and tape them closed. The goal is to create a slider that fits loosely around the skewer. It should easily slide back and forth without getting stuck. When in doubt, make it loose. With an awl, poke holes in two corners on the short side of D1 and D2, ¼" (6 mm) from the edges. Insert the skewers into one of the pieces and attach with glue **(figs. c and d)**.

3. Double check the looseness of your rolled tubes by rotating the skewers and letting the tubes fall freely. If they don't slide all of the way down with ease, they're too tight **(fig. e)**.

4. With a knife or scissors, cut a craft stick into two 1½" (3.8 cm) sections and glue them to either side of the rolled tubes. Make sure you glue them in a way that still allows them to slide along the skewers **(fig. f)**.

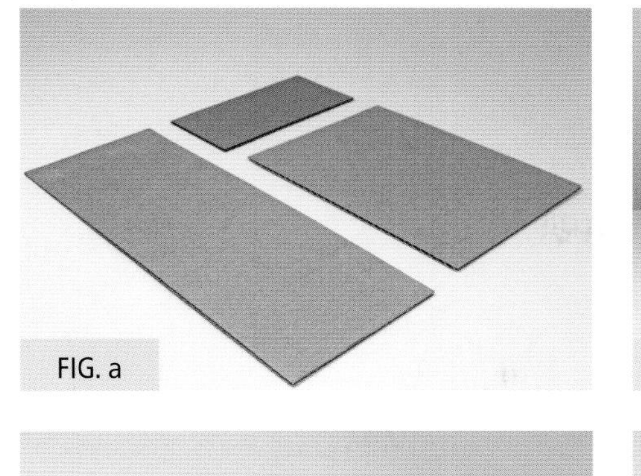
FIG. a

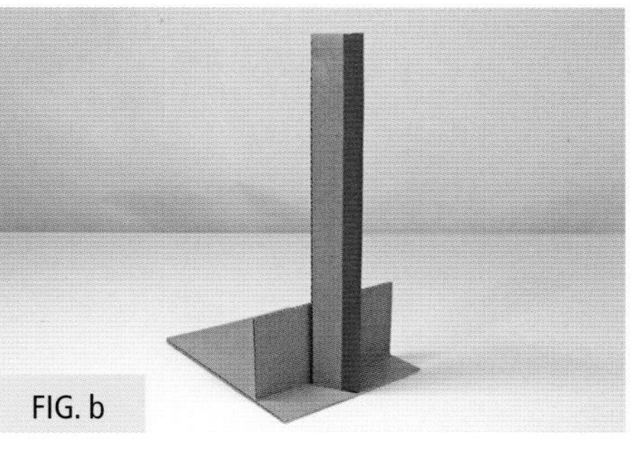
FIG. b

FIG. c

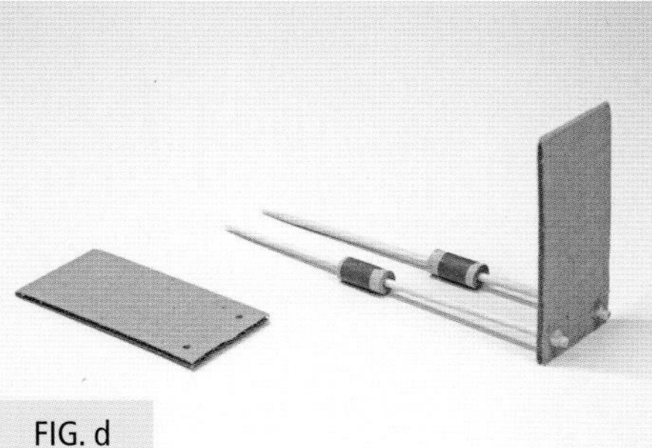
FIG. d

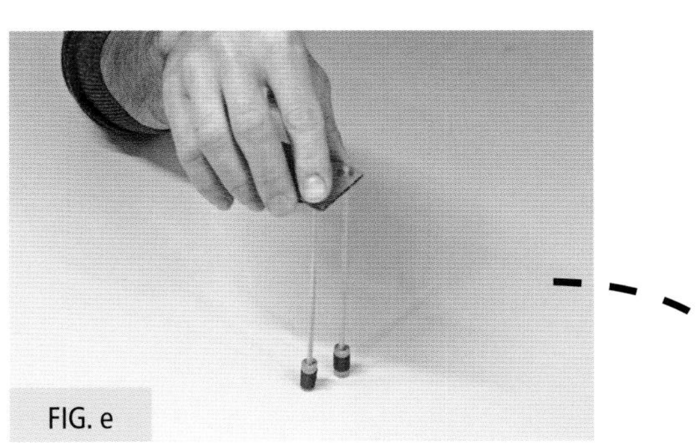
FIG. e

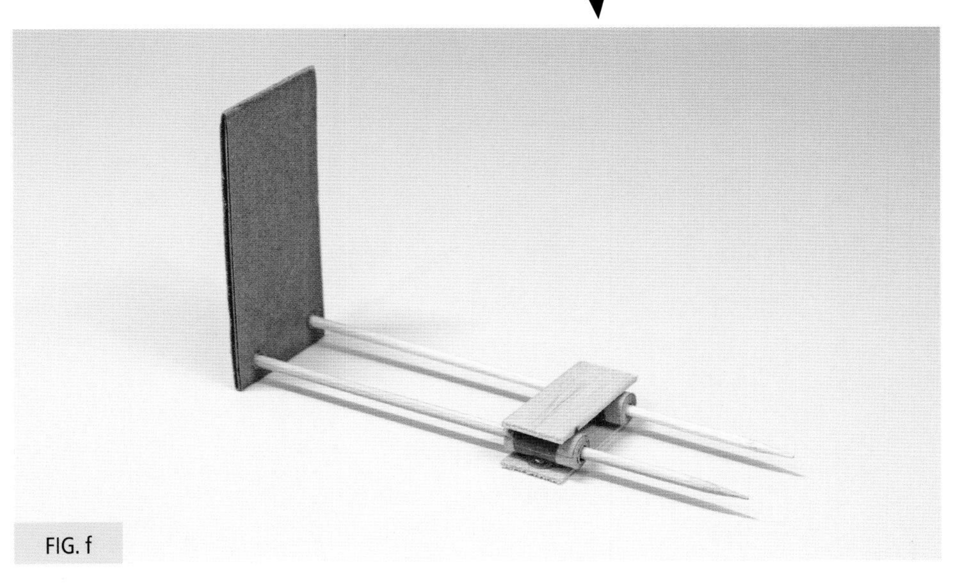
FIG. f

5. Glue the skewer structure to the triangular upright 1½" (3.8 cm) from the top, and add C1 for support **(fig. g)**.

6. Slide on the rolled tube piece and attach the bottom to the triangular upright. Add C2 to mirror the top **(fig. h)**.

7. Cut the rounded ends off of a craft stick, and then cut it into three pieces, 1¼", 2", and 2" (3.1, 5, and 5 cm). Split the 1¼" (3.1 cm) piece lengthwise into three sections **(fig. i)**.

8. Attach these pieces to a craft stick as shown to make the piece that will hold your cookie. Trim the rounded bit off the top of the craft stick **(fig. j)**.

9. Glue this piece to the rolled tube slider part. You might need to trim the cardboard support so it doesn't interfere with the craft stick as it slides up and down **(fig. k)**.

10. Glue a 1" (2.5 cm) piece of skewer to the top of the slider piece **(fig. l)**.

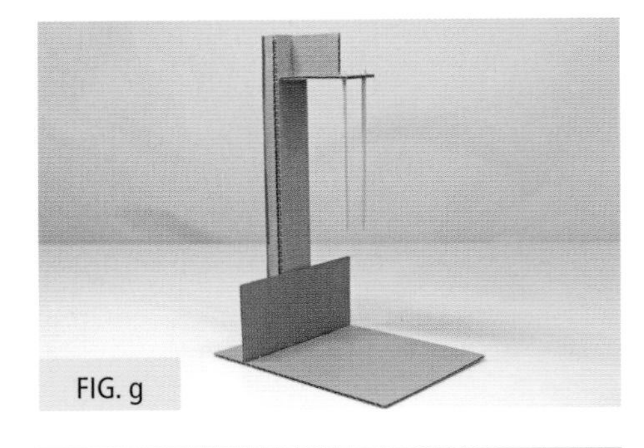

FIG. g

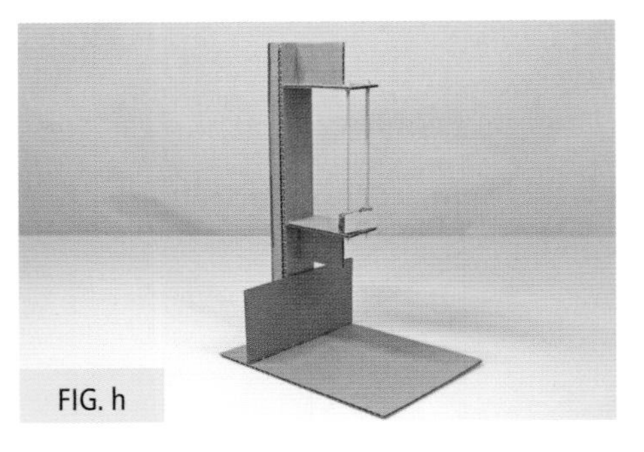

FIG. h

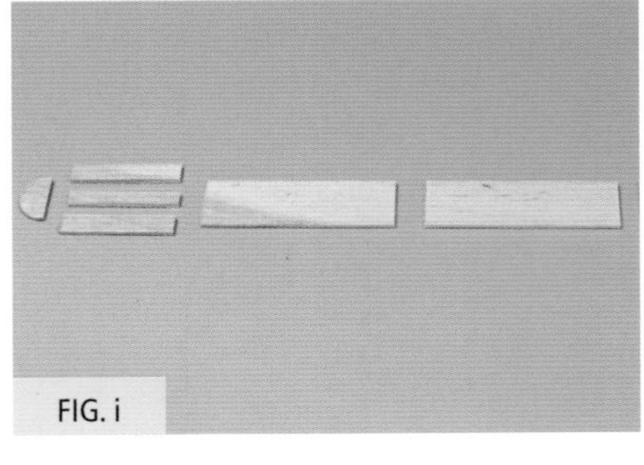

FIG. i

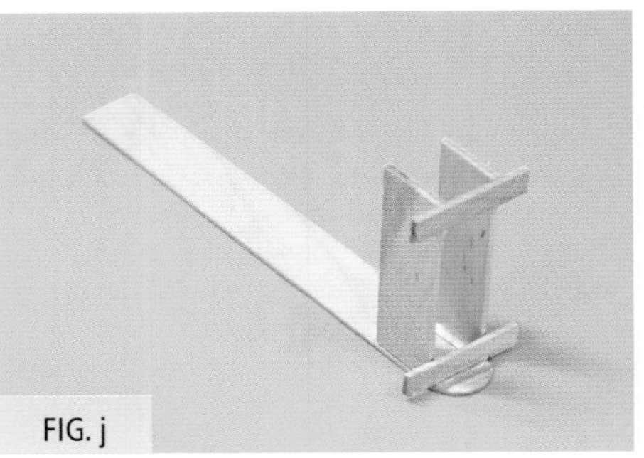

FIG. j

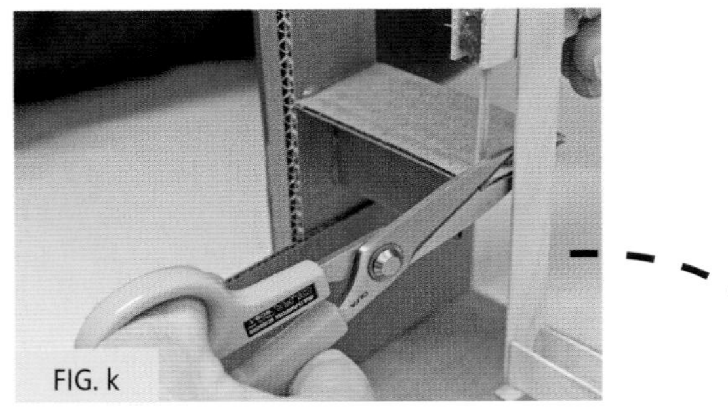

FIG. k

FIG. l

## Make the Lever

1. Stack S1, S2, and S3, and secure together with tape. Poke holes in each end ½" (1.3 cm) from the edge. Poke holes in one corner of Q1 and Q2 and in the center of B1 and B2 **(fig. m)**.

2. Push a 1" (2.5 cm) piece of skewer through the end of the S lever and put B1 and B2 on either side. Then slide on Q1 and Q2 on either side and glue to K for a base. Add B3 to the front of L **(fig. n)**.

## Make the Tower Structure and Linkage Mechanism

1. Score lines 2" (5 cm) apart on J to create a triangular tube (see page 14 for how to score) **(fig. o)**.

2. Find the center of O1 and O2 (see page 12) and poke a hole. Create a cylinder using O1, O2 and M (see page 17) **(fig. p)**.

3. Find the center of N1 and N2 (see page 12) and poke a hole. Attach N1 and N2 to both sides of O1 and O2, and carefully enlarge the hole in the center with a sharpened pencil. Remove the sharpened pencil and replace it with a pencil with no point. Secure this with glue **(fig. q)**.

4. Poke a hole all the way through the tower structure about 2½" (6.4 cm) from the bottom and enlarge with a sharpened pencil. Use B4 as a washer around the pencil on the outside. Add glue to the pencil and outside washer to keep it in place. You can also wrap the pencil with tape to keep the washer in place **(fig. r)**.

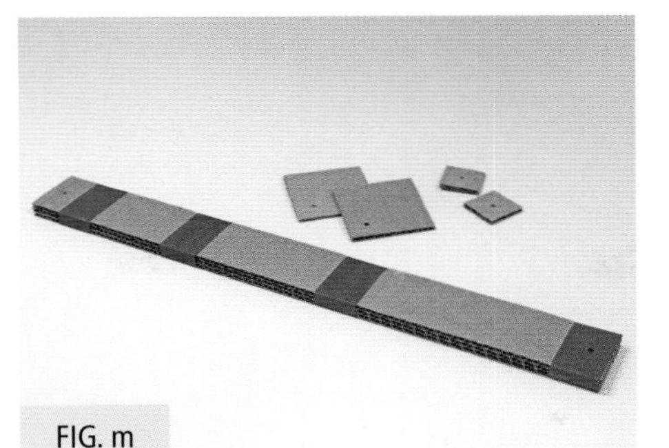
FIG. m

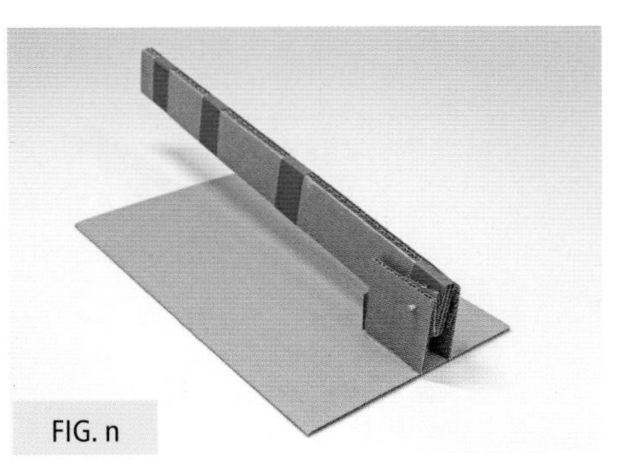
FIG. n

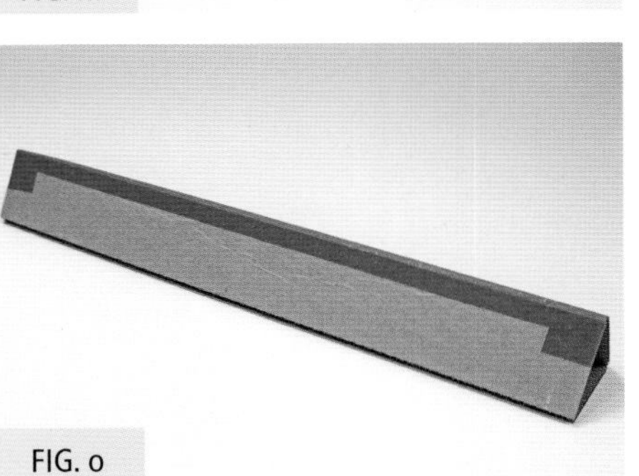
FIG. o

FIG. p

FIG. q

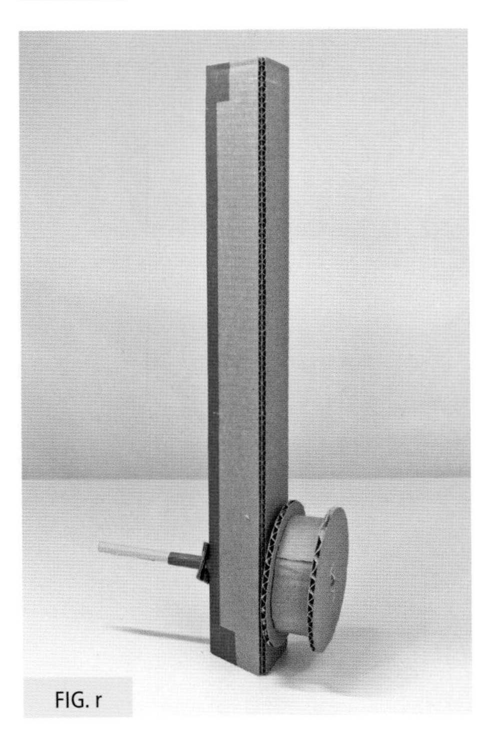
FIG. r

5. Glue the tower to the lever base so that the cylinder is centered over the lever and is about ¾" (1.8 cm) from the Q supports. Add L for extra support **(fig. s)**.

6. Glue together P1 and P2 to create a short arm. Poke a hole ¼" (6 mm) from each end large enough for a skewer to rotate freely inside of it. Poke a hole in the cylinder ½" (1.3 cm) from the edge. Insert a 1½" (3.8 cm) piece of skewer and attach it permanently with glue. Place one end of the short arm over the skewer and add tape to the skewer to keep the arm in place. The cylinder and the arm should both rotate freely **(fig. t)**.

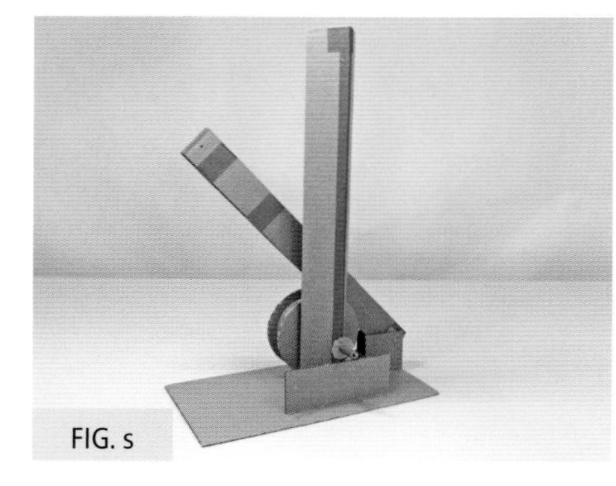
FIG. s

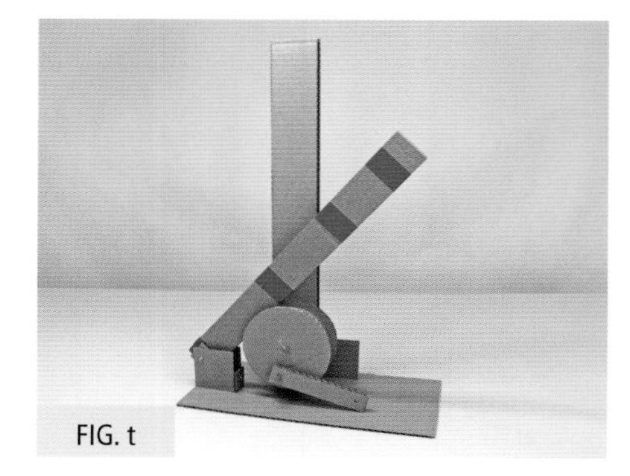
FIG. t

### Make the Lever Tower Structure

1. Score, fold, and tape X1 and X2 to create triangular tubes. Double up Y pieces and glue then to the triangular tubes so they stick out 1½" (3.8 cm) from the top. Poke a hole ¾" (1.8 cm) from the top **(figs. u and v)**.

2. Glue together Z1 and Z2, but only place glue on the outer edges; if you apply glue over the whole surface, you might gunk up the holes and the slit you have to make, and we need those to be smooth in order for the skewers to rotate and slide easily. Poke a hole ½" and 6" (1.3 and 15.2 cm) from the edge. Make a slit 1½" (3.8 cm) long ½" (1.3 cm) from the other edge **(fig. w)**.

3. Place a skewer through the center hole of the long lever and place the uprights on the outside of it. Leave a little gap between the walls of the uprights and the lever, then glue them to the center of AA. Trim the skewer so ¼" (6 mm) or so sticks out on both sides **(fig. x)**.

4. Insert a 1" (2.5 cm) piece of skewer into the short arm attached to the cylinder and attach with glue. Slide the skewer through the hole in the long lever and add tape to keep it in place. Rotate the cylinder until the short arm is vertical when the long lever is horizontal. Make a mark where the small base is in relation to the large base, then glue the two bases together **(fig. y)**.

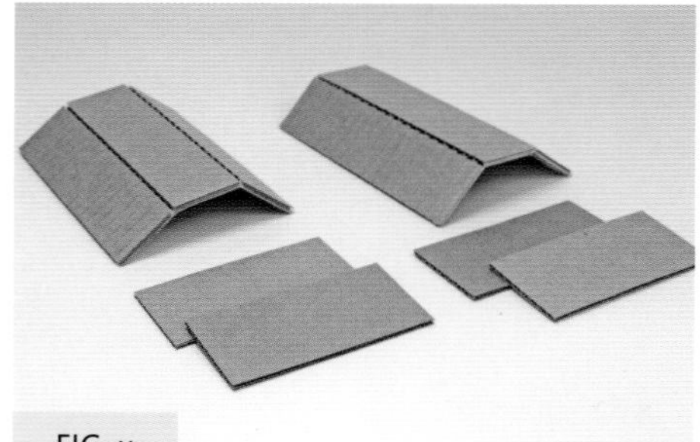

FIG. u

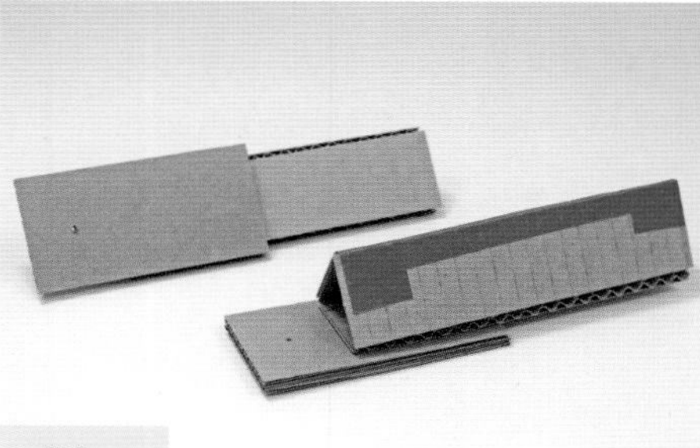
FIG. v

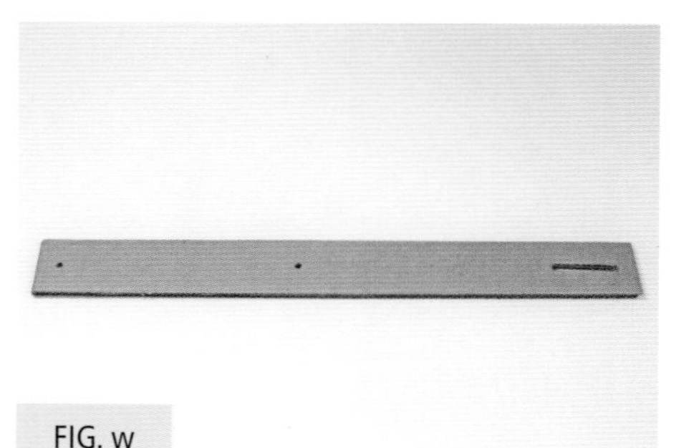
FIG. w

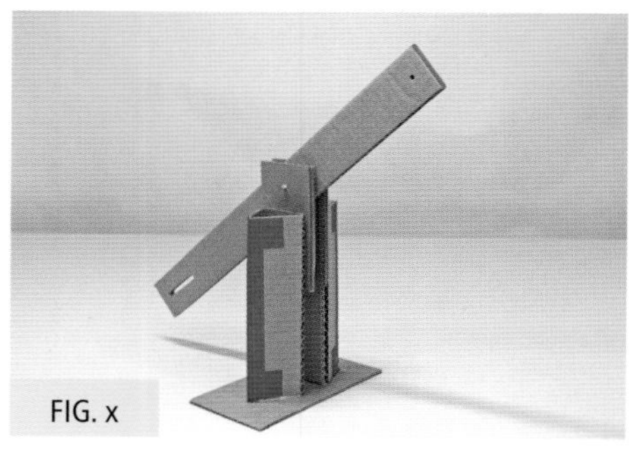
FIG. x

FIG. y

5. Tape the two batteries to the end of the long arm. If you don't have batteries, use anything that's small and heavy, like a stack of coins taped together. If you've covered your hole with tape, repoke it and tie one end of the string through it **(fig. z)**.

6. Poke a hole ½" (1.3 cm) from the top of the triangular tower and insert a skewer **(fig. aa)**.

   Position everything so it matches the photo, and then loop the string over the skewer and attach it to the cylinder with glue and tape.

   Trim the corners off the end of the lever so it doesn't interfere with the supports.

7. Insert the skewer from the dunking mechanism into the 1½" (3.8 cm) slit on the lever. Add tape to the skewer to keep the lever in place **(figs. bb and cc)**.

8. Make a large straw by rolling up R and covering it with tape. Place the balloon around it **(fig. dd)**.

9. Check to make sure your dunking mechanism and cookie fit into the cup you've chosen. If you spin the cylinder by hand, it should raise and lower the cookie. If it gets stuck or is too tight, you might need a bigger cup. Once you get it right, trace the outline of the cup so you know where to put it once you fill it with milk **(fig. ee)**.

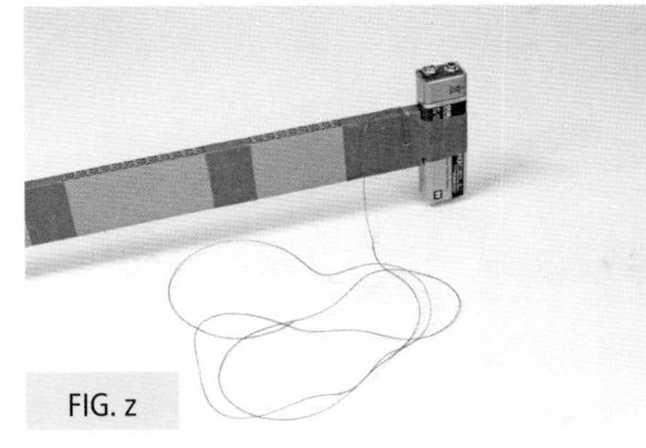
FIG. z

FIG. aa

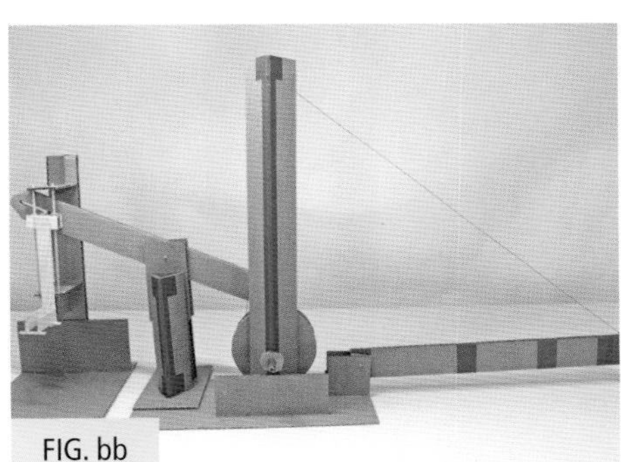
FIG. bb

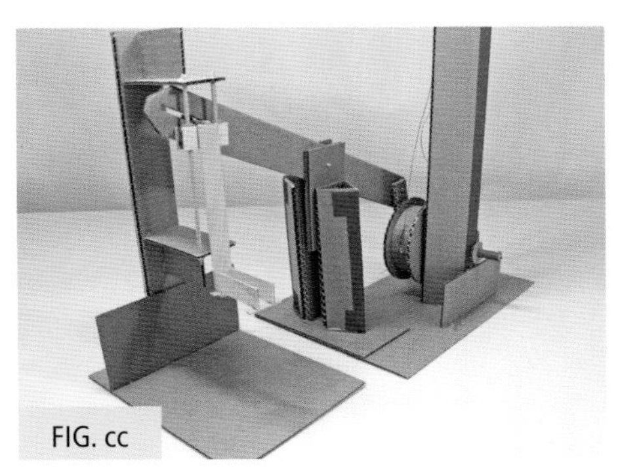
FIG. cc

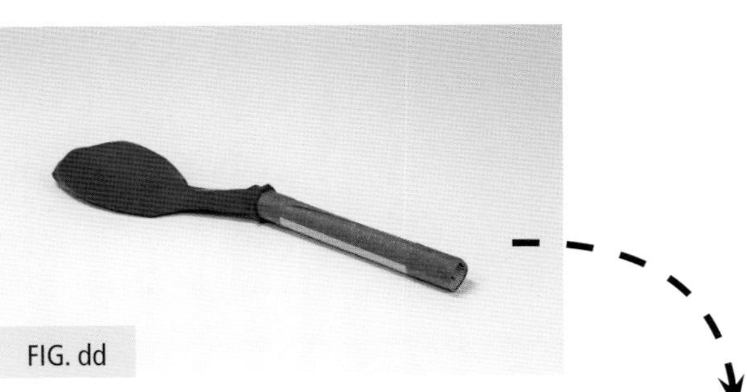
FIG. dd

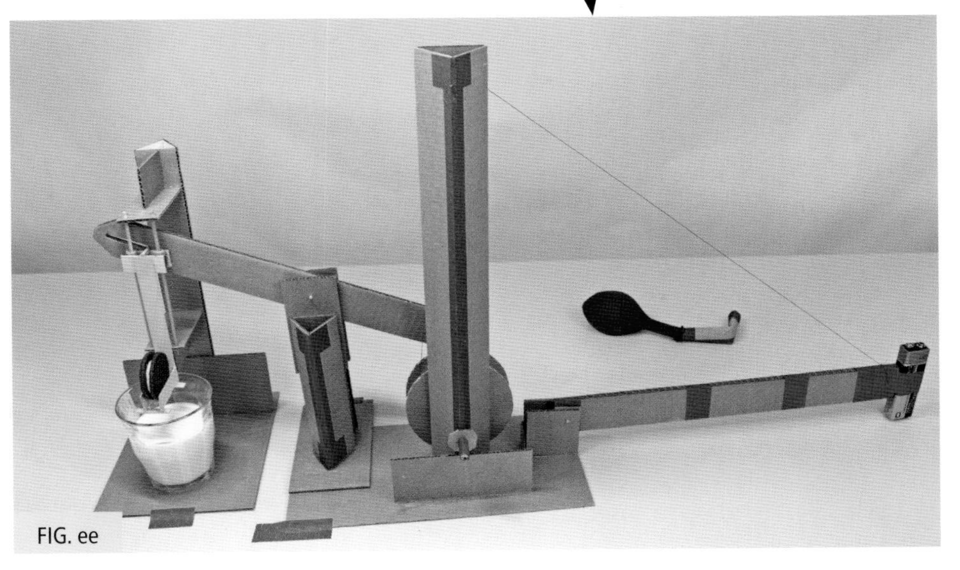
FIG. ee

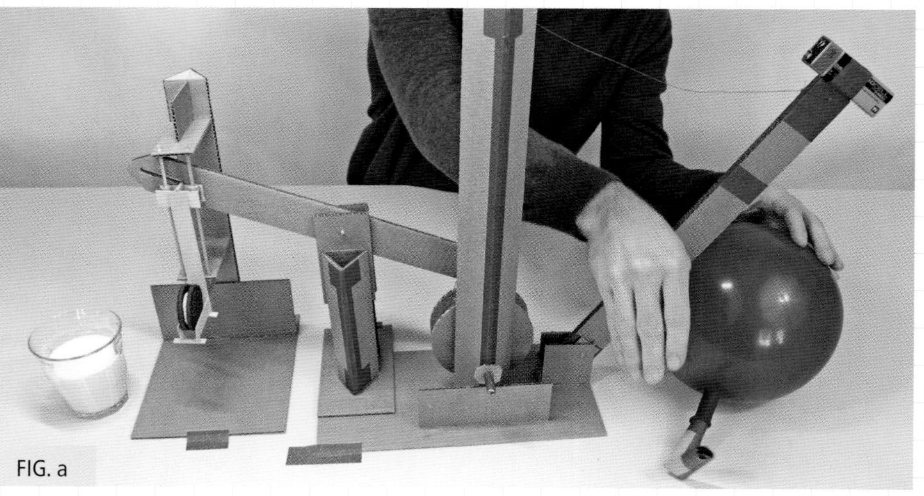

FIG. a

FIG. b

# START THE CHAIN REACTION!

### The Setup

1. Tape the bases of the machines down to your tabletop to secure them. Wind up the cylinder so the string wraps around it. At the same time, this will lift the long arm with the batteries attached to the end. Rotate the cylinder one full rotation so the cookie is in the "up" position. Inflate your balloon by blowing into the paper straw. Crease the straw so the air doesn't come out. Place the balloon under the lever so it keeps it in the upright position **(figs. a and b)**.
2. Before you place the glass full of milk in its place, it's best to do a test run. I discovered that my batteries weren't quite heavy enough, so I added a small bag of pennies. Once everything looks good, slide your cup full of milk under the cookie **(fig. c)**.

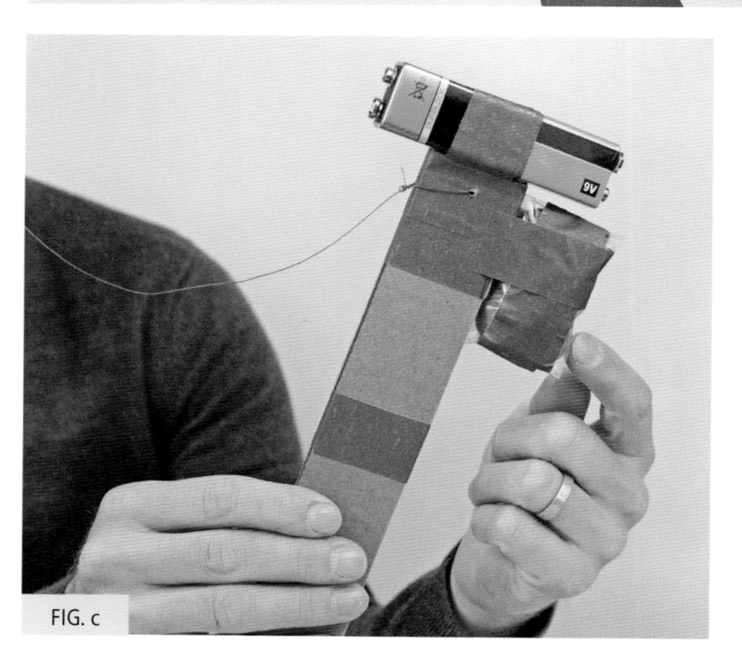


FIG. c

## Make It Go!

To set off the chain reaction, gently straighten the straw so it starts to leak air. As the balloon gets smaller, the long lever starts to fall. As it falls, it rotates the cylinder, which causes the other lever to go down. Once the balloon is fully deflated, the long arm is all the way down, which makes one full rotation of the cylinder, causing the cookie to lift out of the milk.

### Engineering Know-How

This machine converts rotational (or circular) motion into linear (or straight) motion. It does this by attaching a coupling rod to the wheel. As the wheel rotates, the coupling rod pushes up and down on one side of a lever. The other side of the lever is attached to an object that's vertically constrained. That means it can only move in one direction, up and down.

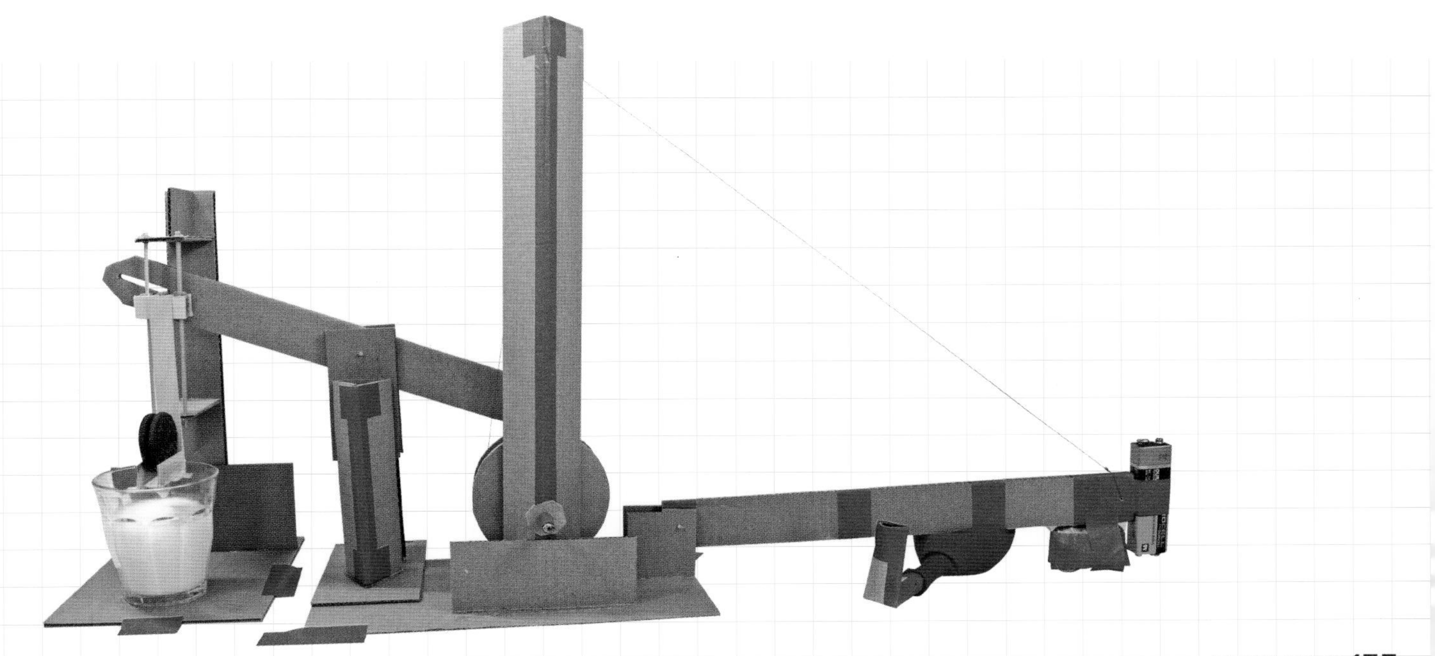

### Troubleshooting

If the cylinder doesn't want to rotate freely, it might be because the hole the pencil is in is too tight or you've accidentally taped it to the upright. Or it might just be that your long arm lever isn't heavy enough, which is what happened to me. Try adding some extra weight to see if that fixes the problem.

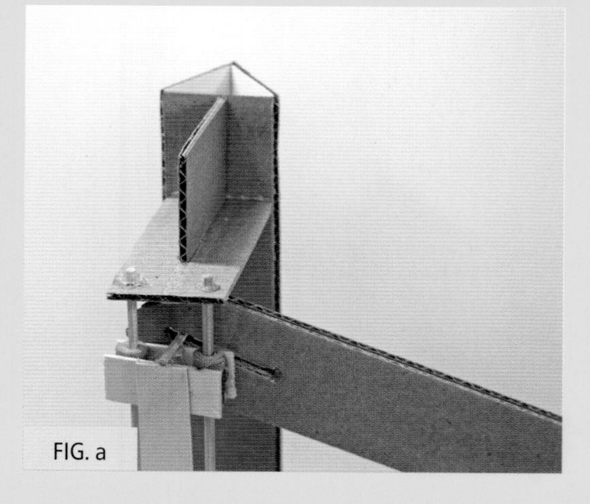
FIG. a

FIG. b

I also had an issue with my cookie dunker not fully lowering. The slit in the arm is as wide as it is so you can move the dunker mechanism from side to side to adjust how high and low it goes. Initially, the shaft on the dunker is on the lower end of the slit. If you untape the base and move the whole thing over a bit until it's at the high end of the slit, you can gain ¾" (1.8 cm) or so between the maximum and minimum height of the cookie **(figs. a and b)**.

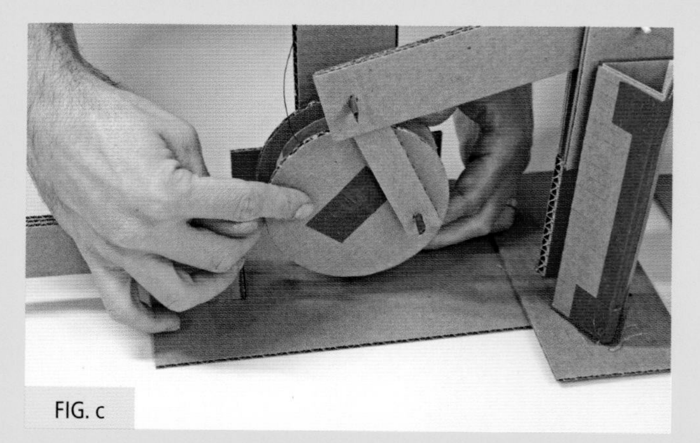
FIG. c

If the pencil or the glue on your cylinder is catching on the short arm, you can add some tape over it to try and smooth it out **(fig. c)**.

FIG. d

If the moving parts are too close together and rubbing, it'll create friction, which might cause the machine to not move freely. Make sure there's a small gap between all moving parts **(fig. d)**.

# RESOURCES

## BOOKS

***The Art of Tinkering***
By Karen Wilkin and Mike Petrich

***Five Hundred and Seven Mechanical Movements***
By Henry T. Brown
Online edition: www.507movements.com

## CREATIVE AND INSPIRING YOUTUBE CHANNELS

**Denha**
Country: Japan
Topics: Marble machines and other gadgets
www.youtube.com/user/denha

**Matthias Wandel**
Country: Canada
Topic: Woodworking
www.youtube.com/user/MatthiasWandel

**Izzy Swan**
Country: United States
Topics: Woodworking and wild contraptions
www.youtube.com/user/rusticman1973

**Bruce Yeany**
Country: United States
Topics: Cool science projects and experiments
www.youtube.com/user/YeanyScience

**Wintergatan**
Country: Sweden
Topics: Machines and homemade musical instruments
www.youtube.com/user/wintergatan2000

## ONLINE LEARNING

**DIY**
Free forum where kids post and view tutorials to acquire new skills
www.DIY.org

**JAM**
Online courses for kids
www.jam.com

**Instructables**
Classes on a variety of topics
www.instructables.com

## OTHER WEBSITES TO DRAW INSPIRATION FROM

**Make:**
www.makezine.com

**Paul Long (my website!)**
www.paulglong.com

**The Rube Goldberg Machine Contest**
www.rubegoldberg.com

# ACKNOWLEDGMENTS

Quarto

- Jonathan Simcosky, for being the first to suggest creating a kids' book.
- Joy Aquilino, for walking me through all of the bits and pieces that go into making a book from scratch.
- Anne Re, for her stellar support on all things art related.
- Karen Levy, for the quick edits on all of my mistakes.
- Meredith Quinn, who might be the fastest email responder out there. She worked tirelessly to get this book right, and helped me slog through all of the edits and tweaks. I think there were close to a million.

JAM

- Chalon Bridges, for helping me figure out how to make a course for kids, which led to this book, and for the long chats to figure out everything else.
- Kelsey Holtaway, for walking me through the course-making process, and pretending I wasn't so terrible on camera.

Mrs. Smith, my elementary school teacher who saw something special in every single kid that walked through her door, and made sure they had the tools and support to figure out what they were good at.

My mom, whose curiosity was embedded in me at an early age. Self-taught in nearly all she does, her love of learning and endless support are inspirations to this day.

My dad, who always brought home old computers and VCRs for me to take apart. Thanks for not getting too mad when you saw the holes I melted in the carpet with the soldering iron, or found bottle rockets in the eave of the house.

And Golbanou, for setting up a photo studio in our tiny, tiny apartment and tirelessly taking every single photo in this book (along with the hundreds that didn't make it inside). The kitchen table was for months covered in bits of cardboard, tape and hot glue, but you always found a way to make it work. Your patience and talent far exceeds mine, and I'm already thinking of all our future collaborations.

# ABOUT THE AUTHOR

**Paul Long** is an engineer and educator. He received his master's degree in mechanical engineering at the University of Louisville. He teaches an inventions course for kids at Jam.com and spends his spare time tinkering with cardboard and sewing the perfect backpack. Paul strives to inspire people to create things for themselves by using random objects to build interactive and kinetic sculptures. He is fascinated with all things moving (especially gears and the wings of birds), and he gets a kick out of combining natural elements with mechanical and man-made items.

Photo by Sarah Hebenstreit

# ALSO AVAILABLE

**Math Games Lab for Kids**
978-1-63159-252-2

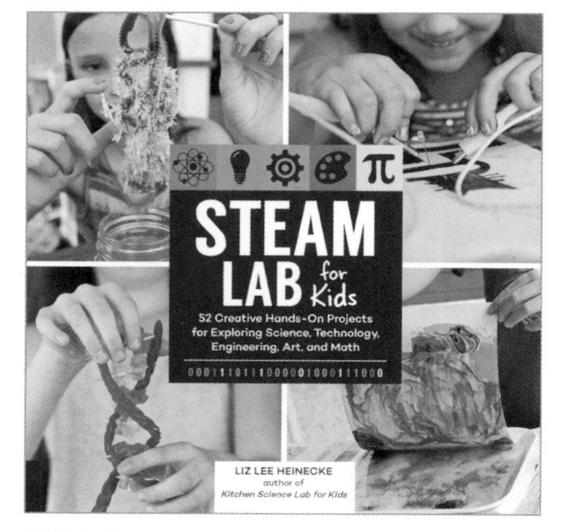


**STEAM Lab for Kids**
978-1-63159-419-9

**Unofficial Minecarft STEM Lab for Kids**
978-1-63159-483-0

**Brain Lab for Kids**
978-1-63159-396-3